AF834196

Undoing the Damage

Silviculture for Ecologists and Environmental Scientists

Undoing the Damage

Silviculture for Ecologists and Environmental Scientists

PAUL A. WOJTKOWSKI

CRC Press
Taylor & Francis Group
Boca Raton London New York

CRC Press is an imprint of the
Taylor & Francis Group, an **informa** business
A SCIENCE PUBLISHERS BOOK

First published 2006 by Science Publishers

Published 2018 by CRC Press
Taylor & Francis Group
6000 Broken Sound Parkway NW, Suite 300
Boca Raton, FL 33487-2742

First issued in hardback 2018

© 2006, Copyright reserved
CRC Press is an imprint of Taylor & Francis Group, an Informa business

No claim to original U.S. Government works

ISBN-13: 978-1-138-46862-7 (hbk)
ISBN-13: 978-1-57808-426-5 (pbk)

This book contains information obtained from authentic and highly regarded sources. Reasonable efforts have been made to publish reliable data and information, but the author and publisher cannot assume responsibility for the validity of all materials or the consequences of their use. The authors and publishers have attempted to trace the copyright holders of all material reproduced in this publication and apologize to copyright holders if permission to publish in this form has not been obtained. If any copyright material has not been acknowledged please write and let us know so we may rectify in any future reprint.

Except as permitted under U.S. Copyright Law, no part of this book may be reprinted, reproduced, transmitted, or utilized in any form by any electronic, mechanical, or other means, now known or hereafter invented, including photocopying, microfilming, and recording, or in any information storage or retrieval system, without written permission from the publishers.

For permission to photocopy or use material electronically from this work, please access www.copyright.com (http://www.copyright.com/) or contact the Copyright Clearance Center, Inc. (CCC), 222 Rosewood Drive, Danvers, MA 01923, 978-750-8400. CCC is a not-for-profit organization that provides licenses and registration for a variety of users. For organizations that have been granted a photocopy license by the CCC, a separate system of payment has been arranged.

Trademark Notice: Product or corporate names may be trademarks or registered trademarks, and are used only for identification and explanation without intent to infringe.

Visit the Taylor & Francis Web site at
http://www.taylorandfrancis.com

and the CRC Press Web site at
http://www.crcpress.com

Library of Congress Cataloging-in-Publication Data

Wojtkowski, Paul A. (Paul Anthony), 1947-
Undoing the damage : silviculture for ecologists and environmental scientists/Paul A. Wojtkowski.
p.cm.
Includes bibliographical references (p.).
ISBN 1-57808-426-1
1. Forests and forestry--Environmental aspects. 2. Forest management. I. Title.

SD387.E58W63 2006
634.9'5--dc22

2006042327

The greatest impediment to scientific innovation is usually a conceptional lack, not a factual lack.

(Gould, 1989)

Silviculture can never be fully learned from books, nor in the lecture room, and success in practical forestry can only be attained by observation of the manner in which forest trees and crops grow, and of the influence of local conditions of soils and climate.

(Gol, 1910)

Preface

As a natural science, silviculture has a large say in how humans interact with the terrestrial world. Although the perspective taken here, the production of wood, is narrow, the amount of land area consumed is extensive; the indirect consequences of wood production on natural processes are larger still.

Through the amount of land engaged, the flora and fauna affected and the environmental consequences, good or bad, silviculture is a frequent constituent in applied ecology, environmental science, conservation ecology and other broad land-use disciplines. Silvicultural expertize is essential when trees and wood are an economic output; best promoted when silviculture is allied with hydrology, ecology, soil science, wildlife management and other conservation sciences.

At present, silviculture has an ad hoc relationship with the other land-use disciplines. More can be done, both to bring ideas in and to integrate silviculture to an environmental whole. The direction is clear, the recasting of silviculture as an offshoot of ecology is replete with the appropriate underpinnings.

As taught and practiced, conventional silviculture centers around bio-simplicity (as shown by the commonness of, and continued emphasis on, the plantation monoculture). There are many stories, past and present, of practitioners taking simplicity to the extreme; cutting down natural forests, replacing these with tree monocultures. The drift should be in the opposite direction.

The lack of multi-species plantations, and the failure to actively promote these, is a severe shortcoming of the current system. Under ecological tutelage, bio-complex plantations become less foreign and more mainstream.

In defense, the push for simplicity did not always prevail. Silviculturists once placed great stock in biodiversity. The opinion expressed here is that practitioners erred over 100 years ago. Instead of seeking shelter in simplicity, it is along the biodiversity front that silviculture becomes fully relevant. This early misdirection still guides the field.

Conventional agronomy, without agroecological input, suffers the same tendencies and drawbacks. Monocropping dominates the productive process while environmentally-harsh chemicals replace beneficial natural forces. Meanwhile, natural influences, those that can be directed toward increasing yields, remain terribly underutilized. These shortcomings and the potential solutions are shared with silviculture.

With wood production as the key economic or use rational, the road that connects ecology with silviculture runs through agroecology. Foremost among the gains are a strong theoretical and conceptional base. With underlying theories and principles, silviculture moves from a discipline based on trial-and-error to one of greater embodiment and exploration.

One does not have to look far for potential add-ons. Among these, nature, rather than people, has a greater role in managing silvicultural entities. And, in addition, native flora and fauna thrive if provided with a welcoming, bio-diverse setting.

In addition to the environmental blessings, revenue increases and/or cost savings make biodiversity a sensible business practice. The purpose for changing the underpinnings of silviculture is to shed old ideas, breakdown traditional barriers and initiate this new (or renewed) bio-direction.

Acknowledgments

For their invaluable assistance, the author would like to thank the libraries at the University of Massachusetts at Amherst; Harvard Forest; Williams College; Berkshire Community College; Cornell University; and the Berkshire Athenaeum at Pittsfield, MA. The author also thanks Professor Matthew Kelty, University of Massachusetts at Amherst, for his help with some of the references and the Massachusetts Department of Environmental Management (Massachusetts DEM) for several of the photos.

Not all contributors can be named. It goes with saying that, over a long career, countless coworkers, colleagues, hosts, and other individuals have shared their thoughts or have helped in gathering such. Hence, recognition and thanks goes to those that toil in obscurity, knowingly and unknowingly contributing to the advancement of sustainable land-use practices.

Contents

CHAPTER 1

Introduction

Silviculture is a long-established field and central to forestry as a discipline. As a human-induced, land-used activity with broad ramifications, silviculture qualifies for inclusion under the headings of environmental and conservation science. If undertaken with safeguards and a deference for nature, silviculture can also be offered as an offshoot of natural terrestrial ecology.

Silviculture exists to provide wood in continuing supply. Wood can be utilized as veneer, lumber, poles, dimension stock, cants, chips or other physical forms. Trees can also be utilized for fiber, as a chemical base stock or as firewood. Saleable and serviceable wood is the economic engine that propels silviculture.

Unavoidably, trees must be destructively harvested to obtain wood. However, if rightly managed, the originating ecosystem remains or can be reestablished.

Any number of land-use practices can continuously supply the end product. Wood-producing, land-use ecosystems range from tree plantations to managed natural forests. Other wood-producing options intercede, some clearly inclusive in forestry, others are at the fringe. Some of the latter might be better inscribed under agriculture, rather than forestry.

In silviculture, sometimes lesser, sometimes higher goals intercede and must be incorporated into land-use decisions. These include wildlife conservation, hunting and/or foraging by local people, tourism, climatic "resilience", the preservation of native flora and water management.

Whatever the case, silviculture often accommodates diverse interests; the driving force being the economic encouragement afforded from harvesting and selling trees. How the singlemost important goal is managed (in this text, the wood output) and how diverse objectives are handled (in this context, these are the environmental and the people-oriented aspects) defines silviculture.

SILVICULTURE

In the history of silviculture, a number of definitions have been proposed. Green (1908) has defined silviculture as the growing of trees in groups and forests. Hawley (1921), SAF (1944), and others have a two-part definition, "the art of producing and tending a forest" and "the application of the knowledge of silvics in the treatment of a forest."

More recent reckonings, and the definition utilized here, have silviculture as the cultivation of forest trees, most often as a wood source. In a more explicit form, as a management system of forest areas as characterized by the typical stand structure, spatial arrangement, management sequence and the felling system (Simon, 2001).

Under a less technical, more conventional perception, silviculture is the raising of trees where wood is the primary or the sole output and only highly secondary co-economic activities are tolerated. Examples where wood is the primary or the sole crop are natural forests managed for wood production, commercial wood plantations and, at the edge of forestry, farm forestry (i.e., blocks of trees in a farm setting, but distinct and apart from agriculture).

Fringe Activities

Relegated to the operational fringe is agroforestry; the mixing of common agricultural crops with trees. In many agroforestry systems, wood is the primary output or of equal value to the accompanying crops. These wood-first versions of agroforestry are the margin of, but squarely within, silviculture and forestry. This is true in the earliest texts, e.g., Evelyn (1664), and continues to the present.

Outside the Fringe

Discussion brings on the question as to where silviculture stops. There are tree-based systems that are clearly agricultural in nature, but at times, with significant wood output.

Treecrops have a primary, non-woody product, but may yield marketeable wood. The consummate case is walnut with both a

Photo 1.1 The two, initially outward faces of silviculture, a managed natural forest and a forest tree plantation (poplar trees with grazing).

valuable nut and desirable wood. After a long stint producing nuts, fruits, spices, latex (as with rubber trees) or some other product, tree crops can become a wood source. Not qualifying as agroforestry, orchards and treecrop plantations fall into a gray area between forestry and agriculture.

Wood, as a secondary agroforestry product, challenges even the more far-reaching silvicultural definitions. Forest trees can be present in agriculture, be an ecologically active part of an agroecosystem, overlay large areas and be a commercial source of wood. Examples are parkland systems (scattered trees in open pastures). Whether these qualify is a topic for debate, most foresters draw a line, placing these within the sphere of agriculture.

For discussion in this text, silviculture encompasses wood-only systems or those land-use practices where wood is the primary output. Through agroforestry, it can be raised in conjunction with substantive agricultural activity.

Outside the scope of this text, but deserving notice, is where wood is grown but it is secondary to crops, forage or some other non-woody output. A glimpse of these land-use options may be warranted, but not such as to advocate a silvicultural definition which extends outside that of forestry.

Although the bounds are not precise, silviculture, as a land-use discipline, should not, or need not, overlap into other environmental realms. The wood producing function occupies a gap along the human-nature interface and an otherwise unfilled niche within the broad spectrum of ecological or environmental sciences. The practical challenge is in satisfying a demand for tree growth and/or harvest strategies; all without inflicting environmental damage or, better yet, as a vehicle for environmental improvement.

Responsibilities

Although wood production is the principal goal, silviculturists are stewards of the land and bear obligation to protect natural flora and fauna, maintain the integrity of the land (at local and regional levels), and to seek a sustainable outcome from prolonged or infrequent use. The focus and intensity of this responsibility depends on the type of silviculture practised.

With large natural forests, stewardship is unconditional, not to the degree where complete protection is provided, but to the point where

(See caption-next page)

Photo 1.2 Three systems at the silvicultural fringe, (a) treecrop (in this case, a rubber tree) where wood is often harvested after a long stint producing non-woody output; (b) a parkland system with scattered trees over pasture, and (c) coffee plantation with a forest tree overstory.

the ecosystem, not the individual plants, is secured. Protection, in a strict interpretation, bars any intervention that is non-natural (such as hunting and logging). This degree of protection can, at times, run counter to the natural natural processes being protected. An untouched, fully protected forest, one that does not provide any economic or useful return, may not be in the best interests of the local people.

With agroforestry or farm forestry, the responsibility shifts more toward maintaining the future productive potential of the land. There is still an imperative to do one's best in maintaining natural flora and fauna and insure other ecologically positive outcomes. However, there is more leeway in the process.

In either case, silviculturists have the responsibility to care for natural ecosystems and to maintain the overall integrity of the land. Knowledge, planning and correct implementation can go far in insuring that wood outputs are profitably obtained. Knowledge and planning also insures that the land and ecosystems are not compromised to the point where all or some of the productive potential is lost. Also inclusive is the requirement that nature does not suffer unduly.

AGROECOLOGY

In this age of ecology, silvicultural goals are better served if placed under an ecological umbrella. As a vehicle for promoting productive silviculture, natural ecology is not entirely appropriate; this approach does not advocate nor support much of the interface between nature and people-based, land-use activities (silviculture included).

For this, agroecology comes to the rescue. Agroecology taps into rich veins of ecological knowledge while plugging the gap that exists between the study of untouched natural ecosystems and those impacted or planned by mankind.

Agroecology, rather than natural ecology, addresses the need to produce food, fuel, and fiber to local, and not so local, communities. As such, agroecology bridges the gap between nature and people and promotes a smooth transition, spatially and temporally, between what nature provides (as site resources, e.g., land, climate, natural vegetation) and what people need (i.e., food, fuel and fiber).

When agroecology, as manifested through environmentally friendly land-use practice, is not present, the nature-people interface can widen. The gap is broadest when native species are threatened, or worse, become extinct. A wide gap is also unmistakable when the productive potential of the land is reduced or lost.

If full advantage is taken of the agroecological land-use possibilities, this gap can shrink. In some cultures, even those without formal agroecology, a respect for natural processes and accommodation of natural dynamics can produce an economically useful landscape, one in relative harmony with nature.

There are a number of facets to a nature-silvicultural interface. The first of these bears on stewardship and responsibility (as mentioned above). The second, a bedrock of the agroecological approach, is in using natural ecosystem dynamics (rather than high levels of labor and/or chemical inputs) to achieve productive goals. This may not be an outright substitute, but one of reduction and replacement. To achieve this, the cues come from nature, specifically how nature, without human interference, accomplishes such tasks. Numerous examples throughout this text show how natural dynamics increase yields (i.e., tree growth rates) and/or reduce management costs.

AGROECOLOGICAL GROUPINGS

With deference to and a clarification on, the definitional discussion of silviculture, agroecology subdivides into five categories (Wojtkowski, 2002):

(1) agronomy,
(2) agronomic agroforestry,
(3) integrated agroforestry,
(4) silvicultural agroforestry and
(5) silviculture.

The overall relationship between the academic disciplines and the variants of agroecology are presented in Fig. 1.1. Three of these, (1) integrated agroforestry, crops on an equal footing with trees; (2)

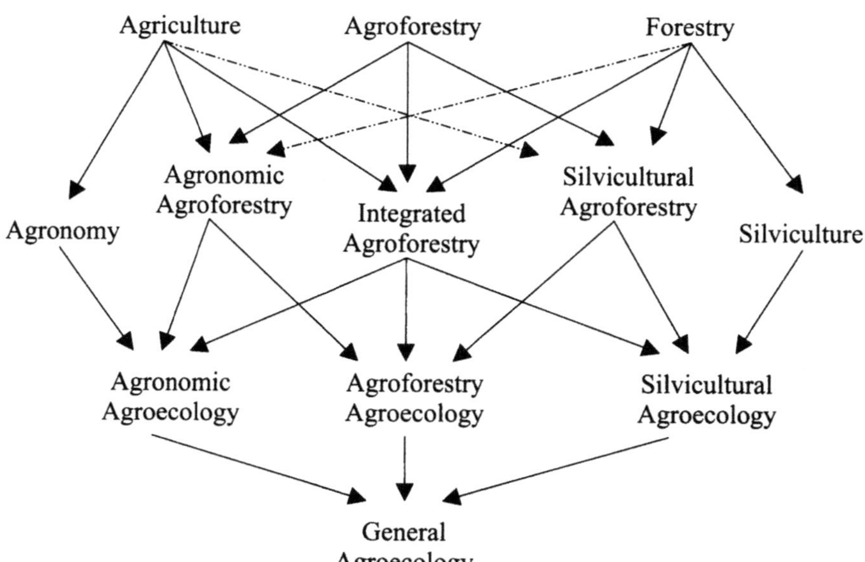

Fig 1.1 The relationship between agriculture, agroforestry, and forestry and general agroecology. The unequivocal and intermediate branches of forestry include silvicultural agroecology with subheadings of silviculture and silvicultural agroforestry. Agronomic agroforestry (connected with a dotted arrow) has a ambiguous relationship with forestry. The same holds true with agriculture and silvicultural agroforestry (also identified with a dotted arrow)

silvicultural agroforestry, ecosystems where trees are the primary output, crops are secondary; (3) silvicultural agroecology, wood as the only output; are embraced in this book.

A fourth, that of agronomic agrofoestry, where wood output is secondary to an agriculture output, is commonly perceived as being outside the bounds of forestry. As such, there is an unsettled connection with forestry (represented as a dotted arrow in Fig. 1.1). Correspondingly, crop-producing ecosystems, those where wood is the primarily output, are generally viewed as being outside agriculture. Therefore, wood-first, silvicultural agroforestry has a similarly ambiguous connection with agriculture (the dotted arrow in Fig. 1.1).

Silvicultural Agroecology

Redefining silviculture as an agroecological science helps (1) by placing this within the larger body of ecological thought and (2) bring to the fore many less explored options. In the latter category, the strength of agroecology lies with multi-species systems, many of which have received scant notice under a non-agroecological silviculture. Also on this list are highly promising variations of the monoculture.

Silvicultural Agroforestry

The view espoused here is one where agroforestry, specifically those systems where wood is the primary output, is part of silviculture. This spans and even increases across, generations, e.g., Gent (1681), Browne (1832), Schenck (1904) and Evans (1992). Despite continued discussion, this view has lacked the vigor needed to widen current silvicultural perceptions. As with multi-tree plantations, the mixing of crops and trees is best explained through agroecology.

Integrated Agroforestry

Wood can be equal in value to other marketeable and non-marketeable commodities and some agroecosystems produce a wide range of outputs. Some agroforestry plantations fall under this heading. Equally possible are species-rich, tree-with-crop agroecosystems.

Among the possibilities are agroforests. Agroforests are fully manmade, highly biodiverse, ecologically-sound forest ecosystems, usually with a large percentage of fruiting species (for a full explanation, see Chapter 13). These have undisputedly positive environmental properties and, in the right markets and locations, profitably produce wood. Common in many regions, these champion the notion that agricultural output (as with marketeable fruit) can enhance the economic attractiveness of wood-yielding ecosystems.

SILVICULTURAL SUB-DIVISIONS

There are other sub-divisions within forestry and silviculture. Prominent among these are the common tree categories: (a) coniferous, (b) broadleaf, (c) deciduous, (d) evergreen, (e) hardwood, and (f) softwood.

Although discussion along these lines is not germane to this text, these serve to illustrate the problems (1) of classification, (2) of terminology, and (3) of natural variation that is fraught with exceptions. The hardwood and softwood categorization shows the weakness of any one grouping. Balsa (South America) and the baobab tree (Africa), both with the softest of woods, are commonly perceived as hardwoods.

The same problems pervade forest types and ecosystem classifications. Among the divisions are: (a) humid and dryland; (b) tropical and temperate; (c) ordered and disarrayed; and (d) natural and planned. These are also fraught with exceptions (as is much of ecology) and serve best as descriptors when in combination.

Featured here are two other classifications, that of (e) gap and patch derived ecosystems and (f) species and ecosystem governance. The first of these, derivation, goes to the dynamic origin of most forest ecosystems. From this, arguments are made (Chapter 14) that support responsible silvicultural as an ecologically compatible activity. The second, governance, looks at the ecological governing mechanisms for agroecosystems, e.g., tree plantations, a common silvicultural expression, are species governed.

This is a bit more involved than a silviculture subdivided into only species-simple plantations and species-complex natural forests. The other divisions are useful in understanding and formulating systems that, on a regional level, utilize natural dynamics to achieve yields and revenue, reduce costs and risk and address other management goals.

THE NATURE–SILVICULTURE INTERFACE

How silviculture treatments are applied, both to maximize wood yield and to maintain a deference with nature, is far from a simple question. Ecosystems can be far more dynamic and far more accommodating of change than commonly perceived.

A lot depends upon what is the forest norm for a given area. Although this line of discussion is in flux, authors and articles eschew the notion that forests, without human intervention, mature into dark, brooding, closed-canopy, entities where change comes only at a slow pace.

In most cases, forests are dynamic, open ecosystems that should be managed as they evolved e.g., Attiwill (1994), Coates and Burton, (1997), Wohlgmuth et al., (2002). With dynamic ecosystems, there exists considerable latitude to find harvest prescriptions for natural forests and to formulate tree plantations that promote what nature intends. This justifies well-contrived silviculture as a nature-compatible activity.

COMMUNITY FORESTRY

How a local community interacts with forests and silvicultural activities affects the type of silviculture practised. This can be with the species planted, types of management undertaken (if any), and the harvesting and processing methods.

In some regions, silviculture can be distant, without much direct influence, but important in daily life (e.g., as a source of clean water). For others, there is a direct community association where many earn their livelihood from the forest. Silviculture can also be integrated, through farm forestry or agroforestry, into agriculture where trees are essential, economically and/or ecologically, toward a favorable farm outcome. Whatsoever be the case, the community, broadly defined, has an input, directly or indirectly, into the type of silviculture practised. This is the silviculture–community interface.

THE UNFOLDING OF SILVICULTURE

It helps, at least initially, to view silviculture as has having two-outward faces. These are the planning and management of (1) forest-tree plantations and (2) natural forests. Ecological governance, at the ecologically fundamental level, fronts these two lines of development.

Systems of low biodiversity (e.g., most forest-tree plantations) tend to be species-governed, agroecosystems of high biodiversity (e.g., most natural forests and some species-rich plantations) tend to be ecosystem governed.

Previewing and overviewing what lies ahead, Fig. 1.2 shows the unfolding of these two development tracks. Starting with the agrobionomic principles (and an explanation of governance), various implementation and applied use concepts fill the conceptional space between pure theory and field practice. In a further gain, this progression also refines, sharpens and interprets the argument that silviculture is a subset of agroecology.

At the intermediate levels of biodiversity, species-rich plantations and forests share, with simpler ecosystems, sundry management

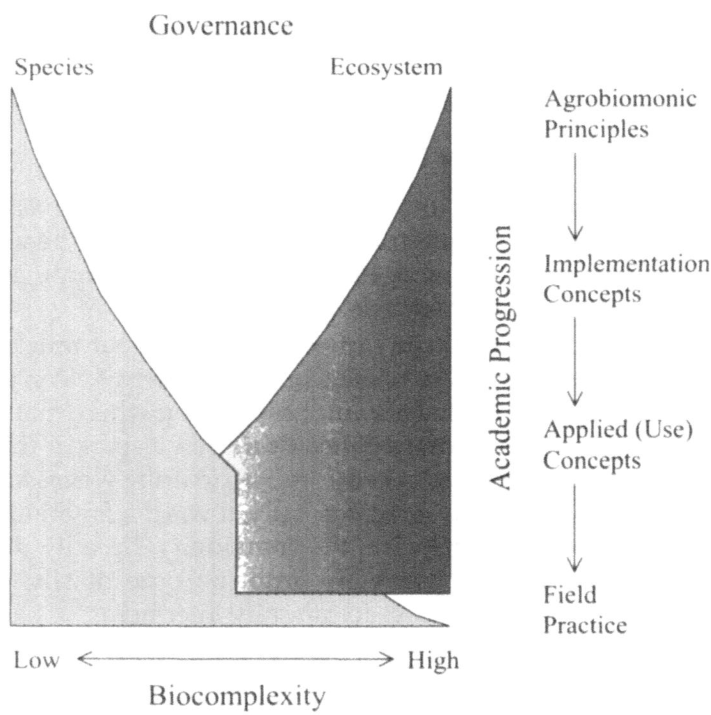

Fig. 1.2 Previewing the layout of this text, the progression proceeds downward from basic theory to field practice with intermediate linking steps. The two vertical tracks represent forms of governance, either species or ecosystem. Chapter 2 begins this progression, delving into the principles of governance.

techniques, risk containment measures, and, where applicable, sustainability concerns. Accordingly, the two development tracks overlap (as shown in Fig. 1.2) as undiluted theory makes way for the nuances of practical application.

CHAPTER 2

Agrobionomic Principles

The agrobionomic principles are foremost in the development progression of this text. On a conceptional basis, these principles help connect natural ecology to agroecology and on to the silvicultural version. Starting with the niche theory, these explain ecological happenings within agroecosystems, the end goal being the design of environmentally friendly, productive ecosystems that benefit landusers, consumers of wood products, and, with enlightened design, nature.

NICHES

In order to survive in a highly competitive world, species (whether plant or animal) utilize or occupy different niches. These define the outward being of a species and its relationship with the living and non-living world it comes into contact with.

For plants, these niches are complex and varied. Table 2.1 samples a few of the plant niches, vis-à-vis the site and other plants, that allow for survival and growth. These can be singular or function in combination to overcome the various stresses present in a plant's immediate environment.

Much of this, and the importance of the niche, come to the fore with those resources essential for growth. This lies in the capture and utilization of light, water and the principal nutrients (nitrogen, potassium, and phosphorous) along with a host of trace elements (e.g., iron, copper, zinc, calcium, etc.).

Table 2.1 can only hint at what is a complex array of species attributes and responses. Nature is messy and convoluted. Things seldom fall into neat categories; exceptions and/or unique cases are not hard to find. This includes some eccentric plants that have evolved

Table 2.1 A few of the niche characteristics (properties and conditions) that differentiate plants

Properties
Root locations (density, level and/or extent or vertical and horizontal spread)
Timing of
foliage bloom and loss (includes leaf loss for winter or dry/wet seasons)
growth (optimum and sub-optimal periods)
Resistance to various destructive microorganisms (e.g., diseases)
Adaptation to positive microorganisms (e.g., in-soil mycorrhizae)
Adjustment or resistance to fauna (e.g., insects and grazing animals)
Seed dispersal (e.g., wind, water, or fauna assisted)
Wind resistance (intensity and/or duration)
Water accommodation (high and low rainfall levels, soil moisture content and duration of the event)
Fire resistance (intensity and/or duration)
Relative height and growth rate
Growth needs
Soils (includes maximum, minimum and optimal conditions)
nutrient content (N, P, and K plus trace minerals)
pH
moisture (surface and/or sub-soil)
bulk density
depth
Temperature (max., min. and optimal)
growth (optimal) and survival (max. and min.)
germination (most or some plants may lack an optimal)
Humidity (often combined with soil moisture and temperature)
Light
direction (vertical or horizontal)
duration (e.g. day length)
intensity (latitude or cloud related)
frequencies (portion of the light spectrum utilized)

for unique growing conditions or unusual competitive advantages (Emboden, 1974).

One result is a multiplicity of opinions on what defines or what is the role of a niche (Liebold, 1995). The discussion here takes a more pragmatic approach. Under the agrobionomic principles (this chapter), niches serve as a base from which to convert generalist concepts (as covered in this chapter) into use-specific practices.

Fundamental Niche

The fundamental niche occurs when a plant species lives sans competition. This helps explain what may be expected when a plant enters the world of the competitive realized niche. A single plant growing in a spacious flowerpot occupies a pure and unqualified fundamental niche.

Inclusive in the fundamental niche are the plant attributes that promote growth on a site. A conspicuous example is the white or paper birch. The white, smooth bark is a cold weather adaptation that allows an unshielded tree to survive, without splitting, the combination of extreme cold and strong sunlight (the white bark reflects unwanted heat).

Of silvicultural concern is stem and wooa quality. A fundamental plant characteristic is the ability for producing a clear bole without competition. Eucalyptus and American elm (see Photo 2.2) are species where this is inherent. Other species require some form of competition for this to occur.

A variation of the shared or constrained fundamental niche has application to, and ecologically defines, the monoculture. This is where multi-species relationships (a realized niche) are absent, but a plant must contend with competition from niche-identical or niche-similar species. Although some of the principles described in this chapter do apply, fundamental niches ecologically detach the monoculture from other silvicultural forms.

Realized Niche

A realized niche is that which exists with interplant competition. This can be with a one-on-one, inter-species partnership or that which exists within a complex, multi-species ecosystem. Some plant species are content to live within the confines of an ecosystem simply by exploiting the essential resources not used by others. This may be termed a passive niche. Other species are not so acquiescent.

Aggressive growth can elicit essential resources in the face of interplant competition. Examples are pioneer trees where, through rapid propagation and fast initial growth, these purloin available resources after some calamity eliminates or greatly reduces competition. Plants that follow the aggressive species are often passive, existing on unused resources.

Some species are even more aggressive, exhibiting fast growth but, through high population densities, deny other species room to grow. This property is found in cover crops where high density prohibits later arrivals from establishing.

Some plants force themselves into an ecosystem through physical and chemical means. Physical force is less common, but does exist. Strangler vines exploit this strategy, growing on established trees to reach sunlight. At this point, they develop a self-supporting stem, strangle their unintended host, taking space and essential resources from their temporary, and doomed, supporting tree.

Some plants extract more than their share of an essential resource and, by denying others, keep competition in check. Species of eucalyptus are thought to couple fast growth with the overuse of water. The competitive niche, that which excluded other species, functions best when water is the limiting resource (as in moisture constraining situations).

Soil chemical properties can be imposed by one species to keep others in check. Pines prefer acid soils, redwoods prefer alkaline soils and, by actively maintaining soils in this state, this discourages other plants.

Chemical competition can be employed to clear away competition. Through allelopathy, chemical compounds are produced by some plants to the detriment of others. The mechanisms, as well as the types of compounds, vary. Some inhibit nearby seeds from germinating, others interfere with growth. Allelopathy may be more common in drier regions where, in seeking water, over competition is detrimental to all concerned.

The aggressive nature and the mechanisms of competition resisting trees have application in agroecology. A large segment of silviculture, especially low intensity monocultures, rests upon plant-on-plant, especially tree-on-weed, aggression.

More appropriate to agroecology are the mechanisms of cooperation. In natural ecosystems, plants have co-evolved and passively share realized niches. It is the harmonic aspects of co-evolution that, through biodiversity, offers unexplored potential for economic gain.

SUCCESS IN APPLICATION

The fundamental measure of success in mixed species systems is the land equivalent ratio (LER). This is

$$LER = (Y_{ab}/Y_a) + (Y_{ba}/Y_b) \quad (2.1)$$

where Y_a and Y_b are the yields from monoculture plots of species *a* and *b*. The yield of species *a* grown in combination with species *b* is Y_{ab} and the yield of species *b* with species *a* is Y_{ba}. In silviculture, the above yields are expressed as an annual growth increment, seasonal change in diameter, or some other comparative measure of wood volume.

For this, an equivalent ratio of 1.0 indicates that the growth performance (and the utilization of essential resources) is in line with that of the monoculture. The higher the value, the greater the indication that site resources (and the site itself) are being engaged with ecological efficiency. A LER less than unity denotes that excess plant-on-plant competition is undermining efficient resource utilization, i.e., plant growth is suffering because of an inability to amiably coexist on a site with the resources available.

AGROBIONOMIC PRINCIPLES

The agrobionomic principles stem from niche theory. The myriad of concepts presented under this heading are divided into two systems of governance: (1) species and (2) ecosystem.

Species-based governance establishes the operating principles or mechanisms for bicultures, three-plus polycultures, and other bio-complex, mostly ordered systems. Ecosystem-based governance underlies the management of natural forests, agroforests and other bio-rich ecosystems.

SPECIES GOVERNANCE

Species governance relies upon one-on-one dynamics including the exploitation, vis-à-vis the niche characteristics of the individual species, of available essential resources. Ideally, niche relationships are exploited through specific species pairings and plant-plant interrelationships. Much of this involves the division of essential resources between competing species. The sub-divisions of species governance are:

Competitive Production
- (1) acquisition
- (2) partitioning

Facilitation

Exclusion
- (1) partial
- (2) full

Competitive Production

Competitive production occurs when two or more productive species cohabit the same site and experience better growth and yields and/or increases in the overall worth of the outputs. The concept of overall gain (with multiple and dissimilar outputs) is defined by the LER. The mechanisms of competitive production are:

- (1) competitive acquisition (interception gains);
  - (a) time or season of essential resource utilization is different,
  - (b) the resources are utilized in the same time or season, but
    - (i) with different collection zone,
      e.g., plant root profiles are in different strata, and/or
    - (ii) resources are derived from different sources,
      e.g., N_2 capture from the air by legumes vs. in-soil NO_2 acquisition by non-legumes.
- (2) competitive partitioning (conversion gains);
  - (a) one resource is used more efficiently by two species,
  - (b) two species, through different resource-use profiles, utilize a combination of resources more efficiently, and/or
  - (c) resource removal by one species can, for the second species, increase the
    - (i) overall productivity,
    - (ii) the harvest index, and/or
    - (iii) the quality of the output.

Photo 2.1 shows two species exhibiting, through close proximity, some degree of success in competitive production. Although clearly coexisting, the exact mechanisms that underwrite coexistence can be hard to determine.

Photo 2.1 Close proximity and good growth indicate successful competitive production is occurring. In this case, wild cherry grows close to black walnut.

Acquisition

On any one site, there may be more essential resource available than a single species can acquire. Sunlight striking the ground or reflecting off leaves is one such instance. When two or more species intercept a greater percentage of an available and limiting essential resource, there is potential for greater overall yields.

Palm (1995) has put inter-plant capture (nutrients moving between species) at only 20%. This may be indicative of amount available and the relative amount captured. Minerals remain unavailable and untouched because they are part of living biomass (micro and macro) or chemically bound or physically trapped in unyielding rocks. Others may simply be overlooked.

Temporal

In silviculture, opportunity exists in pairing species by their temporal procurement of resources. The common case is where one species is more active earlier in a growing season, the second may come to the fore later in the season after the first has finished a growth spurt and has a diminished need for site resources.

As a visual indication, many deciduous oaks take on leaves late in the growing season after other species have established theirs. This growth strategy may take passive advantage of late-season water and nutrient availability.

Zones

Sunlight is among the essential resources that have zones of interception. Trees with broad, horizontal canopies capture vertical, midday light. In contrast, there are species with narrow, vertical canopies that seek horizontal, early morning or late afternoon light.

Many desert dwelling acacias have thin horizontal canopies. Horizontal, light seeking trees are the silver fir, eastern red cedar and Italian cypress. Photo 2.2 shows two such species, an American elm (horizontal canopy) and a clove tree (vertical canopy).

Another exploitable zone lies with rooting patterns. Some species rely more upon surface roots, others survive with roots in deeper stratum. In a Hawaiian dry forest, the tree species *Metrosideros polymorpha* and *Reyoldsia sandwicensis* capture sub-surface water through deep roots whereas the species *Diopyros sandwicensis* and

Photo 2.2 Two canopy types: a horizontal canopy (American elm) that gathers mostly vertical midday light, a vertical canopy (clove tree) that gathers horizontal morning and afternoon light. Note that clearer stems can be associated with horizontal light collection.

Nestegis sandwicensis seek moisture from light rains through surface roots (Stratton et al, 2000).

A commercial forestry example is paulownia *(Paulownia elongata)* which has 80% of the roots 40 to 100 cms below the surface (Wang and Shogren, 1992). As a result, this species intercrops well with shallow-rooted plants.

Some species have the ability to change rooting patterns in response to competition (Weaver, 1919 and 1920; Schroth, 1995). There are indications that some oaks can change their root posture to accommodate competition (Büttner and Leuschner, 1994; Leuschner et al., 2001). This is a case where a fundamental niche can be quite different from the realized niche.

Figure 2.1 shows the two favorable root relationships. The upper shows vertical separation, the lower has roots in different soil horizons.

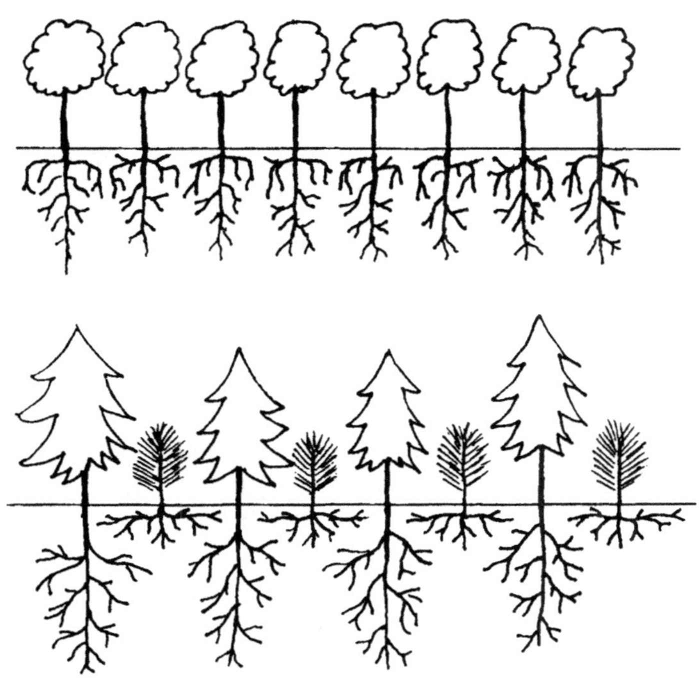

Fig. 2.1 Different forms of root separation. The upper figure shows vertical stratification which can occur with one or many species. The lower illustration shows horizontal separation between unlike species.

Different origins

Species can have similar roots and canopies, but if these species find minerals from different origins, acquisition gains occur. The classic example is those legumes that capture atmospheric nitrogen (N_2) and non-legumes that exist on available in-soil nitrogen (NO_2).

Plants utilizing different light frequencies are documented less, This may be the case with some tropical understory species, those in perpetual shade, that survive off the light as reflected from the leaf surface of other species.

Conversion

When multiple species utilize the same resources, gains in overall productivity can still occur. As with acquisition, the mechanisms vary and more that one mechanism can come into play at a given time.

Marginal gains

When two or more species seek the same resources at the same time, gains occur if the species utilize the resources more efficiently. This effect is more pronounced when the resources in question are more abundant.

Figure 2.2 shows a yield-resource function for two species. Although each suffers a net loss, the two species grow better because the limiting essential resource is applied more efficiently. In this figure, one half of the resource is given, through competition, to a second species. In doing so, the LER increases from one to 1.60 (i.e., (800/1200) + (650/700) = 1.60).

Multiple marginal gains

The concept of marginal gains can be applied across species (as above) or encompass multiple resources. The latter may be a more accurate rendering of what occurs in complex ecosystems. Figure 2.3 shows the multiple resource situation where what is basically an exchange of resources, results in a better LER.

Figure 2.3 illustrates where, in taking resources (R_1 and R_2) from species *a*, there is only a small loss of yield (Y_a). Giving these resources to species *b*, results in large gain (Y_b). Experimentally, this relationship is hard to confirm. However, the idea is credible and may underlie many successful pairings.

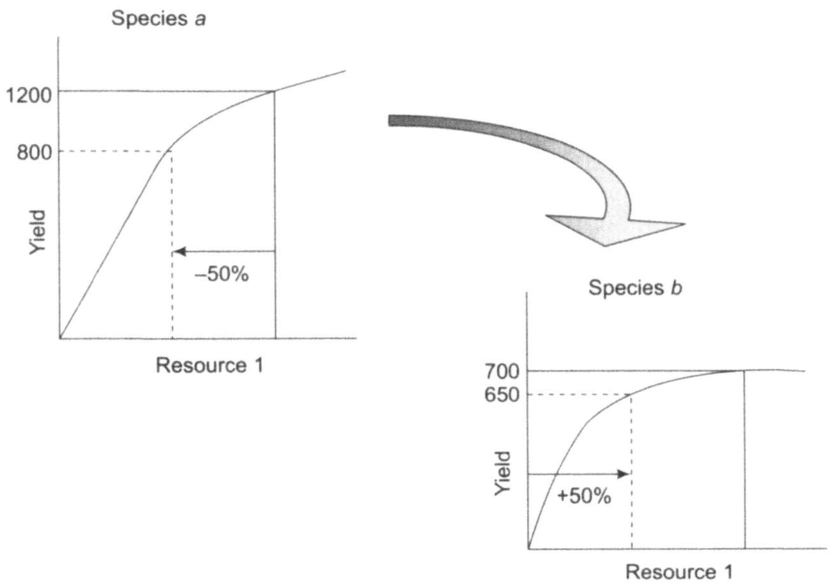

Fig. 2.2 Single resource marginal gains where two species can share the same resource for overall system productive (LER) gain.

Removal

Early writers, e.g., Worlidge (1689), found cases where plants benefit from having fewer resources. Far from an unusual case, examples abound. Water-saturated soil will not favor those trees that are waterlogged intolerant. For some species, high light intensity and low soil moisture causes moisture stress. Resource removal helps in both situations.

For trees, less branching and better stem form is assured if less light is available at the trunk level. In this case, the harvest index (e.g., more usable wood, less canopy or branch structure) and the quality and value of the output can be improved if there is less sunlight at the stem level.

Facilitation

Facilitation is where one species grows or yields better in the presence of another. A double version, i.e., mutualism, is where two species are aided by their mutual presence. Documented examples of single direction facilitation are common and rely upon varying mechanisms.

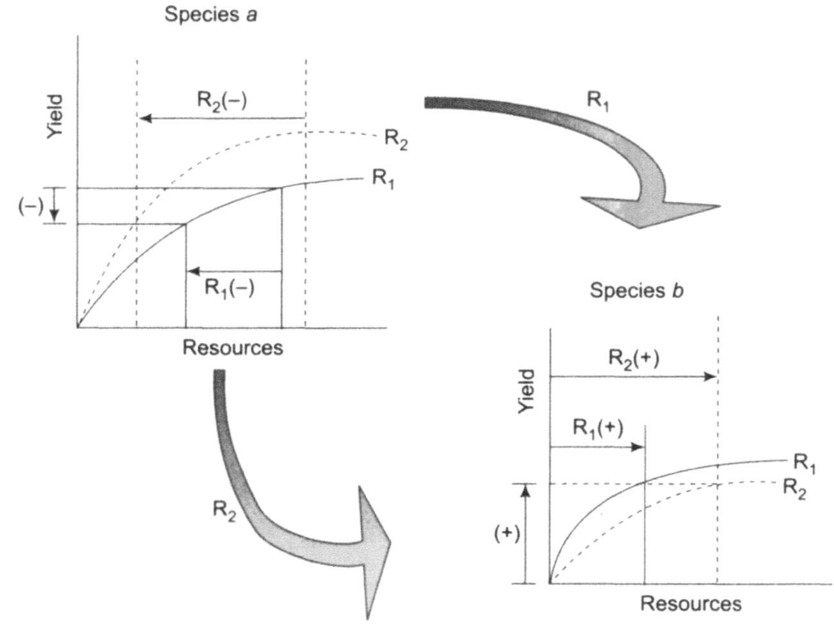

Fig. 2.3 Multiple resource marginal gains where two species with different resource-use profiles share two resources with productive (LER) gains.

By definition, facilitation always results in productivity gains. These can be overall gains or centered on a single species. Again, these can be measured by LER. The categories or mechanisms of facilitation are (from Wojtkowski, 1998):

(1) nutrient capture-transfer
  (a) nutrient pump
  (b) airborne nutrients
  (c) fixation of nitrogen
  (d) chemical conversion
  (e) timing of nutrient release
(2) water capture-transfer
  (a) hydraulic lift
  (b) microclimatic change
(3) retention
(4) accumulation-transfer

(5) concentration
(6) protection
(7) site improvement
(8) parasitic-symbiotic.

Capture-transfer (nutrient)

Nutrient gains are a large part of facilitation and are promoted through varying mechanisms. Some are proven; others are accepted without confirmation.

Among the proven is the fixation of nitrogen by leguminous plants. These enrich the soil, promoting the growth of other species. This is far from a minor effect. Rao and Giller (1993) studied the nitrogen-fixing tree leucaena, finding that 42-54% of the nitrogen found in an understory grass was of tree origin, more if the tree was pruned and the green leaves deposited on the ground.

The capture of airborne nutrients is also an established mechanism of nutrient capture. Some taller plants obtain a share of their nutrient requirement from the air (partially as impacted dust), incorporate this into the plant structure where, upon death and decay, this becomes available to other species. This is also not a minor effect (Szott et al., 1991).

Almost the same occurs belowground. This is where deep-rooted plants obtain nutrients from deep soil strata; bringing these to the surface. These nutrients are converted into plant matter where, through death and decay, these become generally available within an ecosystem. Less supported by research findings, this is still embraced as a positive influence.

The conversion of nutrients is where some species can extract nutrients not available to others. The weathering of rock is a nutrient source and some species surpass others in their ability to get at this bound material.

Under transfer comes the timing of release where, through the rapid decay of organic materials, the availability of the nutrients is accelerated. This may be through the carbon-nitrogen ratio (e.g., Kelty, 1992) and/or more in-humus moisture where the nutrient transfer rate (through decay) is speeded up.

Capture-transfer (water)

The availability of water in ecosystems can be boosted by the addition of a facilitative species. One confirmed mechanism is the hydraulic lift. This is where a deep-rooted species obtains water from a deep source. Plants release this into upper soil layers to dissolve and absorb minerals. In the process, the moisture is preempted by other plants (Dawson, 1993).

Water obtained in deep layers is also transpired from leaves. In the right ecosystem, this can improve the micro-climate, diminishing leaf evaporation and moisture stress in neighboring species. This latter is indirect, but the effect real.

Retention

The runoff of water and the loss of nutrients through erosion and leaching can be averted by appropriate vegetation. Some of the mechanisms of retention focus on water, others on nutrients.

The classic windbreak, in reducing drying, helps retain water. This influence can be quite large, in crops, 20% productivity gains have been recorded (Miller et al., 1973). With some plant species, the ground litter (shed leaves, branches, bark) helps to retain water and can reduce nutrient leaching.

Circulating nutrients, those that escape the attention of surface roots, can be intercepted by deeper roots. These are brought to the surface (through transfer) and eventually become available, through cycling, to other species.

Accumulation

The crop fallow and the accumulation of nutrients is well-established practice in agronomy. Less obvious are the silvicultural applications where one tree captures and holds nutrients until needed by another species. When surplus nutrients are available in the early stages of a plantation (i.e., when the plantation trees are small and widely spaced), these can be retained on-site by a planted covercrop. When the trees grow, die back in the understory releases these to the trees.

The cycling of nitrogen is well recognized and phosphorus can be accumulated in specific plants. The African tree species *Tithonia diversifolia* has been mentioned in this regard (Buresh, 1999).

Concentration

Through root spread, large trees store nutrients gathered at distance. Similarly, the shape of many tree canopies concentrates moisture, funneling water down stems during a high rainfall. Other species shed moisture in a drip zone around and outside the canopy. These mechanisms have application in agronomic agroforestry where crops benefit from stem flow or a drip zone. Pure silvicultural applications are less noted.

Protection

Trees need to be protected from those natural forces that reduce growth or destroy plants. High winds not only reduce available moisture, but can cause leaf and branch rubbing. In crops, this has been shown to reduce yields 20% to almost 50% (Brenner, 1996). Crops, as well as trees, experience breakage and uprooting. In cases of an exposed and poorly designed pine stand, stem breakage was found to be 19-32% and uprooting at 5-9% (Nunishi and Chamshama, 1994).

Protection can be accomplished through internal stand design where less dense, more wind-permeable plantings exert less push on individual trees. Other options include tree stands in wind-protected sites where topography and/or neighboring ecosystems shelter a wind-susceptible stand.

This does not exhaust the protection options. Frost resistance is bestowed when other species block the movement of cold air or reflection heat downward off an overstory canopy. Excessive heat can be countered through the cooling affect of a well-watered, well-ventilated stand.

Site improvement

As with protection, site improvement is a broad category that employs vegetation to alter a number of negative influences. Soil is one area where improvements positively affect the growth of a target-tree species. This can be through temporally induced physical (e.g., decreased bulk density) or chemical properties. The latter includes buffering acidity and alkalinity where subsequent plantings benefit.

As examples, the Indian tree species *Acacia nilotica, Prosopis juliflora,* and *Terminalia arjuna* have been shown to reduce soil alkalinity (Dagar et al., 1995). Other species can reduce acidity and associated problems (Tilki and Fisher, 1998). These can be as monocultural plantings (Lugo, 1997) with the intent of improving the soil structure. The soil will then be right for a later planting of a more valuable, less tolerant tree species.

Other mechanisms are less direct. This can include providing a micro-climate and soil conditions to promote microbial activity and support beneficial micro-flora and micro-fauna.

Parasitic-symbiotic

The host plants for parasites are an unwilling facilitative species. Negative effects are outside the concept of facilitation proposed here, however, this overlaps in the realm of symbiosis where mutual benefits occur.

At the micro-flora level, mycorrihizae help pines in attaining nitrogen. Endophytes, in-plant living fungi, are also inclusive in this rather large grouping. These assist plants in surviving trying conditions, e.g., temperate fluctuations, drought, plant diseases, even reducing the need for sunlight (Milius, 2003b; Pennisi, 2003 and 2004).

Many symbiotic associations are positive, not always toward tree growth, but in economic terms. Truffles are a high-value product found in forest ecosystems and example an economically positive relationship.

Competitive Exclusion

Removing one or more species from competition is the common situation in planned and managed systems. The ubiquitous example is that of weeds. Their removal can direct resources, by default, to the primary species (one or more).

The concepts of exclusion goes beyond this, especially under silvicultural conditions, where some species are kept, but expunged as a competitive threat. These differ from resource removal in that the plant being suppressed contributes no direct productive or economic gain. The categories of competitive exclusion (and suppression) are:

(1) partial (suppression)
  (a) shade
  (b) below ground
  (c) self-suppression
(2) full
  (a) allelopathic
  (b) below ground

These are illustrated in various forms in Figure 2.4. This shows growth rates (left axis) over time (horizontal axis). Growth rates are

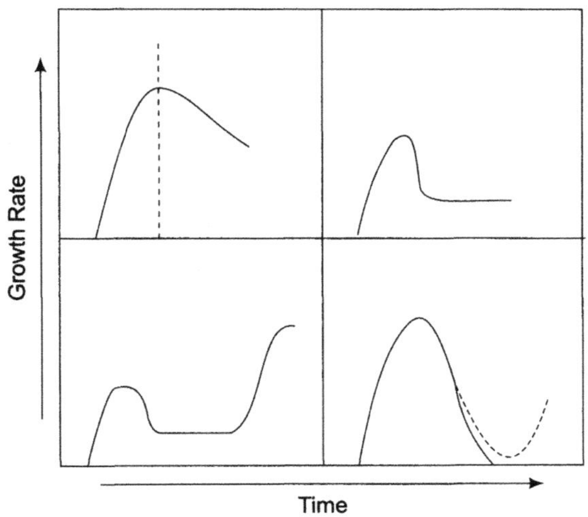

Fig. 2.4 Four suppression cases. Using growth rate (vertical axis) and time (horizontal axis), these are normal, unsuppressed growth (upper left); classic suppression where, after a fast establishment, growth continues at a low level (upper right); suppression where the growth of one species limits that of a second until the first species is removed (lower left); and self-suppression where crowding slows growth and causes many plants to die, the few that live (dotted line) are able to resume higher growth rate (lower right).

the amount of wood added on a yearly basis and this figure shows various exclusionary strategies. Upper left block has normal unimpeded tree growth (Dorado et al., 1997). The upper right block shows strong initial growth followed by partial suppression. Exclusion caused by a second species (lower left), and full exclusion (lower right).

Partial Exclusion

There are times, especially in complex forest ecosystems, where understory (and future succession) is desired, but not to the point where growth of overstory species is compromised. Suppression keeps these species alive with minimal growth while drawing minimal resources.

The mechanisms are shade and below ground competition. These generally work in tandem, but if a situation dictates, these can be separate influences.

Shade

Clearly, a lack of light slows growth. The amount of light is easily determined visually and, with active management, can and does form the basis for maintaining the status quo of the understory. Release comes when the taller species are pruned or harvested.

Below ground

Roots and below ground resource competition are harder to direct but, in a planned sequence, can form the basis for suppression. The key is in denying a species an exclusive below ground niche, forcing competition. This is easily done by surrounding a plant with larger plants of the same species or by surrounding or encircling a taller species with a mix of shrubs and/or other short-statured plants. If the shorter plants have sufficient light, these can out-compete the taller tree for essential below ground resources.

Malcolm (1994) showed that, when horizontal light enters forest fragments, the understory gains at the expense of the overstory. Although not applicable with even-age monocultures, these techniques find use in silvicultural agroforestry when the understory has economic value.

Self-suppression

Counterintuitive in forestry, there are situations where self-suppression can be positively employed. Self-suppression starts with niche-identical or niche-similar plants. To be useful, a series of events must occur: (1) initial fast growth as free resources are utilized, (2) crowding slows growth, (3) many plants die, (4) self-suppression diminishes, (5) the constricted ecosystem is transcended and some plants begin to dominate.

The beauty in this is that, through planting density, the timing of the sequence can be controlled. Seeming inefficient, this may be utilized to control weeds, later producing a marketeable pole crop. A purposeful application of self-suppression, employing densely planted larch and Scotch pine, is given under notable variations, Chapter 10.

Full Exclusion

Removing unwanted weeds is a high-cost activity that is rendered unnecessary with appropriate species dynamics. The methods vary, but effective elimination can be an objective in an overall agroecosystem design. Weeds are only one part of exclusion, facilitative species that have outlived their usefulness are also subject to eradication.

Light

Whereas a little light induces suppression, even less light causes exclusion. A thick compact overstory (one or more) accomplishes this at the ground level. An example of a dynamic sequence is a tree planting with a covercrop. Early, when the trees are young, the covercrop usurps any open niches, eliminating aspiring weeds. As the trees grow, the loss of light gradually eliminates the covercrop, leaving the trees with soil enriched by covercrop residue.

Below ground

As with suppression, a reduction in weeds can be a below ground function. The method, niche denial, employs a mix of species. Although feasible for larger plants (see Chapter 3 - vertical canopy patterns), this is a complicated undertaking for smaller herbaceous species.

Allelopathic

The mechanisms of allelopathy are not well understood, but the ability to repress seed germination is recognized. This has potential for keeping intended plants, while reducing the number of non-intended additions.

Allelopathy can be a property of standing trees, of an established covercrop, or biomass, cut-and-carried to a site. The latter is costly, not often feasible for large-scale forestry, but may prove advantageous in high-input tree nurseries.

ECOSYSTEM GOVERNANCE

In the previous sections, plant-plant species dynamics are explored. Another facet to the agrobionomic principles has the entire ecosystem, not the individual species, as the dominant force.

Ecosystem governance occurs with complex multi-species agroecosystems and the ensuing hodgepodge of inter-species interactions. Some interactions confer strong affect, others are relatively minor. Some cross resource lines, others promote a single essential resource. Some are generated by extending the ecosystem beyond the desired tree species to include shrubs, non-woody vegetation and other organisms, e.g., micro and macro flora and fauna. What results is a summing of natural dynamics where the total exceeds that contributed by the individual species.

Illustrating with a simple ecosystem, water use-efficiency, in capture, retention and conversion, is traceable to a few mechanisms. However, amongst a myriad of biocomplex interactions, there are those mechanisms that are seemingly inconsequential, difficult to observe, and statistically insignificant. Each may be almost negligible but, applied and amplified across all essential resources, these make a system more than the sum of the parts (i.e., the total of the interactions).

Ecosystems

Despite a legion of studies, questions remain as to what constitutes a functioning ecosystem. In all natural systems, delicate balances and less understood niche relationships exist. These can be hard to fathom, but can be alluded to.

Urban streets, backyards and botanical gardens throughout the world contain innumerable exotic trees. Few become integrated in local forest ecosystems, despite growing well in local soils and climates.

Universal examples are pear and apple trees. As full members of the natural ecosystems in the southern Caucasus (Tseplyaev, 1965), these have not been integrated into any forest ecosystem where not native, irrespective of being a long established ecological bystanders.

The American chestnut is another example. Once a dominant species in the eastern forests of North America, an imported disease all but eliminated this tree. Cross breed, disease resistant varieties seem not have the unique combination of niche properties and, despite the now vacant niche, these are not naturally reestablishing (Milius, 2003a).

At the other end of the spectrum are the non-native invasive species that can overwhelm a system, excluding native plants. Rampant exclusion, as when species over-thrive to the detriment of others or when species fail to secure a foothold, demonstrates that the harmonious membership is not an intrinsic ecosystem property.

Membership

Native ecosystems, instead of being a crowd, free for all to join, may be more an exclusive club, open mainly to long-evolved, evolutionary-accredited members. What positions a species with an ecosystem may be slight ascendency, where the species in question gains and maintains competitive niche footholds via-à-vis other plant species. This contrasts slightly with the temporally rigid view of an ecosystem as a niche-based, multi-dimensional jigsaw puzzle where each species has an evolutionary-granted slot.

As long as a species has ascendency, membership guarantees, in spirit, the ability to reproduce within the system and to respect and work with others in maintaining an ecological balance. This is demonstrated when complete ecosystems, not individual plants, overcome adverse soil and site conditions, i.e., conditions that stifle individual species trying to establish without the support of a co-evolved ecosystem (Sollins, 1998).

The exclusive club view is transcended, adding to the debate. Ecosystems of exotics, those coming from far-flung places, do thrive in semi-natural environments (Pearce, 2004). Commonly, these are agroforests (described further in Chapter 13) in which, with a little, or a lot of, outside help, multiple niche-diverse, species can and do have fruitful associations.

The explanation for a successful mix of exotics is that of a trial-and-error assemblage. Those that seek to dominate are, through active management, downgraded to slight ascendency. Those without ascendency are promoted, quietly shelved, or excluded as productive entities.

This topic, that of evolutionary-accredited or trial-and-error membership, can have an impact on or be active concern in silvicultural decision making. Facets of this are developed under thresholds (this chapter), successions (next chapter), and further looked at under complex ecosystems (chapters 12 and 13).

Non-temporal Dynamics

Although ecosystems are temporal entities, they also exist at a moment in time. The ecological dynamics that underlie complex systems do so where, at any given time period, forces are at work that ecologically define a system. This propels them forward in their niche dynamics, ecological development and in their productive role.

Basic parameters

Although ecosystem biological relationships, nutrient and hydrology cycling, energy flows, etc. (i.e., a full complement of cross effects), are complex and difficult to define, these are triggered by three parameters: (1) species density, (2) species biodiversity, and (3) spatial and temporal disarray. These are referred to as biodensity, biodiversity and biodisarray.

There is a fourth parameter, that of duration. Bioduration is the time required for interactions to multiply and spread. As trees take a long time to achieve maturity, this follows the first three parameters.

The agrobionomic principles that underlie species governance do not disappear in a complex ecosystem. These still exist, but are overshadowed or overtaken by the internal dynamics of density, diversity and disarray. In some ways, this greatly simplifies practical management while, at the same time, raising questions on optimal use and design in a productive, bio-complex environment.

Biodensity

Density refers to the number of individual plant species in a given area. Carried to its fullest, this implies some degree of crowding to insure that ecological interactions proliferate (i.e., that most or all niches are fully realized).

Biodiversity

The number of plant species in a given area determines diversity. This can be with many plant species, niche-diverse varieties and plants of uneven ages. To reach a threshold where a fully functioning ecosystem (one having a consequential complement of cross effects) is expected, it is assumed that at least seven different species must be present (Tilman et al., 1997). Ecosystem gains do occur with biodiversity nearer lower limits (Jensen, 1993) and natural ecosystems do well without the full complement of native plants (Schwartz et al., 2000).

The upper limit on biodiversity is quite high. Totals exceeding 150 (Raynor, 1992) up to 300 individual species (Cooper et al., 1996) have been mentioned in planned, managed and species-enriched systems. When launching into this unknown, it may be better to err on the side of greater, rather than lesser bio-complexity.

In silviculture, biodiversity (restricted or expanded) opens sub-parameters and design options. One option is bio-balance, having more or less even number of plants of each species group. Ecosystem governance does require that the number of species exceed the lower threshold.

The other alternative, one more appropriate to plantation silviculture, is to have an uneven population balance. A few dominant, high-value species (as few as 2-4) can be ecologically balanced against a large number of trace species (10 plus) to produce a functioning ecosystem. The latter can be a few scattered, low-value, but ecologically interesting trees (e.g., those that attract wildlife) along with many shrubs and a multitude of ground-hugging plants.

Biodisarray

Biodisarray may be ecologically less important than density and diversity. Still, this helps in promoting ecosystem, rather than individual effects. Disarray has the component species at varying and uneven distances apart. Implemented best through broad bio-complexity, what results are innumerable niche modes, and diversity in the magnitude of the ecological dynamics, all while setting the stage for a wide range of micro-organisms.

An important aspect of biodisarray is spatial evenness (Wilsey and Potvim, 2000). This is where the plant species evenly cover an area, i.e., like species are not clumped. This helps ensure an ecologically positive outcome across an entire ecosystem. Evenness also rules out disarray as being exclusively random (i.e., clumps can be random or non-random, therefore ensuring evenness may require planning). As a topic in plantation design, the forms and theory behind disarray are advanced under spatial concerns (Chapter 3).

Thresholds

A functioning ecosystem, one with full ecological benefits, will occur when certain threshold values of density, diversity and disarray are reached. Research (Baskin, 1994) and field observation show that, above 7-10 separate large plant species, a functioning ecosystem develops. There is debate on the source of the excess interactions, i.e., those that make an ecosystem ecologically greater than the sum of the parts.

A top-down view has the larger plants (trees, vines, shrubs, weeds, etc.) prevailing. This view has the visible biocomplexity, broadened and amplified by the multiplicity of both major and minor interactions, giving the ecosystem its overbearing character. Microorganisms, those supported by a healthy plant community, play a lesser, but meaningful role in the unfolding ecological dynamics, e.g., in decomposing different organic materials (the compounds found in wood, bark, and/or leaves).

Although unheralded in the scientific literature (Klironomos, 2002), microbes, micro-flora and micro-fauna may well dominate ecological dynamics (Reynolds et al., 2003). The bottom-up view has, once threshold conditions merit, microbes as the source of beyond-expected ecosystem dynamics and productivity. In agriculture, the bottom-up view is formalized under the heading of Kyusei nature farming (Gold, 1994). With the complexity and longevity of systems, this certainly spills over into silviculture.

Clearly, microorganisms contribute a lot to tree growth. A large share of this may be species-specific. Birches benefit from microbes that improve soil moisture holding capacity (Statälä and Huhta, 1991). Pines are served by nitrogen-fixing, in-soil, mycorrhiza and by other microbes that make available alternate minerals (Setälä, 2000). Exotic eucalyptus, when inoculated with imported microorganisms, grew better (Garbaye et al., 1988). In the latter case, it can be assumed that the existing micro-flora was more suited to local plants.

Microbial profiles are certainly species-specific and may, as well, be ecosystem-specific. Supporting both premises, Cleveland et al. (2003) found a difference between forest and pasture dwelling microbes.

The bottom-up view does not stop with microorganisms. The ecological interactions encompass and are appended by slightly or substantially larger organisms. These can be plants or animals. The ecological benefits obtained from one, not so tiny, contributor (the earthworm) includes soil enrichment, recycling of organic materials, improved water absorption and host of lesser functions (Hulugalle and Ezumah, 1991; Hauser et al., 1997).

Ants also add to a functioning ecosystem. Risch and Carroll (1982), Stanton and Young (1999) and Molles (1999) show soil enrichment as well as an ability to exclude herbivore insects. Even termites, not often associated with wood production, improve soil structure, water infiltration and holding capacity (Mando and Van Rheenen, 1998). Given the vast number of insect species in some tropical forests, e.g., Rice et al., (1997b), a fair percentage of tiny, not quite microscopic, flora are bound to be tree-specific facilitators, while other organisms are free to provide for the common good.

In silviculture, ecological tasks are not always purely productive (as with increasing tree growth rates). Undertakings, such as weeding, thinning and pruning, are part of silvicultural dynamics, but are outside the capability of most micro-flora and fauna. As a result of this, the primary focus in silviculture must be top-down.

From a growth perspective, there is a lot to be gained promoting, through a functioning ecosystem, the micro-components. There are other associated issues. With the bottom-up view, an ecosystem switches from species to ecosystem governance with fewer tree species than would be the case with a top-down view.

CHAPTER 3

Economic Measures and Spatial Patterns

Governance and the agrobionomic principles provide a base of understanding. Although far-reaching, more is needed to convert basic principles to a field-applicable form. Economic factors and spatial concerns are another step in developing the comprehensive concepts needed for beneficial applications.

For successful silviculture, the importance of economic evaluation is undeniable. For monocultures, a monetary measure is one means to compare systems. This also serves well with more complex ecosystems. However, as complexity builds, other methods offer a more in-depth understanding.

As trees grow, the dynamics change and this must be reconciled. Spatial patterns and pattern arrangements insure the best growth rates over time. There are a number of patterns from which to select, some are ordered, others rooted in disarray.

ECONOMIC MEASURES

As with other commercial enterprises, silviculture is evaluated using economic criteria and any number of measures are possible. In early French forestry, the number of hogs a forest could support was considered a better gauge of forest health and management than wood production (Woolsey and Greeley, 1920). This may not be a bad surrogate, even with updated goals, as this goes to the ecological strength of an ecosystem. In recent time, the most common gauge is wood volume and/or annual growth. These can be applied as is or converted into a monetary unit.

Comparison between ecosystems containing different species and spatial options and the resulting plant-plant complementarity must be assessed. The land equivalent ratio (LER) is a very useful measure of non-economic complementarity. This has been modified for economic use.

Stumpage Value

As trees grow, their commercial value generally increases. This can be a series of value plateaus based upon the diameter of standing trees (stumpage value).

The first plateau occurs when the tree reaches a diameter where it can be sold as poles or pulpwood. The plateau exists because larger pulp logs sell for the same per unit price as smaller pulp logs. The second plateau comes when the trees can be marketed for sawn timber. The highest plateau comes about when trees can be efficiently peeled for plywood.

This stepped value function is illustrated in Figure 3.1. Note that the plateaus are not flat, but reflect that increased volume has additional value.

As markets vary regionally, these plateaus are subject to considerable variation. Regions with active pulp and pole markets,

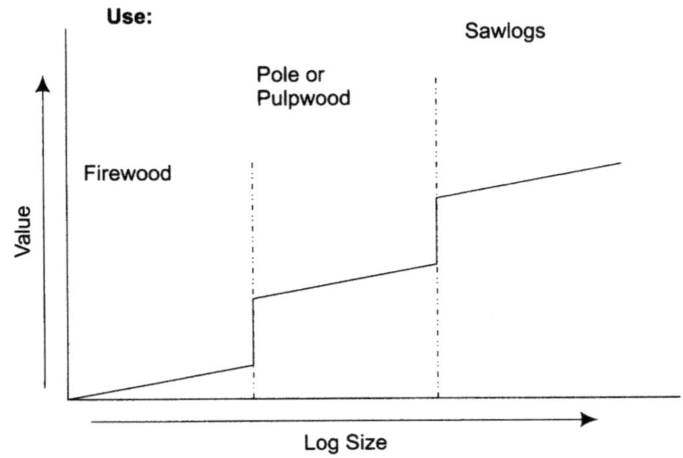

Fig. 3.1 The relationship between log diameter and value. The stepped or plateau function corresponds to the use categories. Those illustrated are for (1) firewood, (2) pole or pulpwood, and (3) sawlogs. The size and height of the sloped plateaus depend on the markets, the region and the species involved.

may not have an outlet for the upper end, higher-valued logs. In some localities, maximum value might be reached in the pole stage. Beyond this the trees are too big to be economically transported and processed with the equipment on hand. In others, especially those far from industry and without cheap to market transportation, buyers will only take the larger diameter, top-quality logs, discounted to reflect transportation cost to processing mills.

There are also quality issues that adjust these plateaus higher or lower. Longer, straighter, knot and defect-free logs command a higher value than those of lesser quality. With this framework, there may be special market needs or adjustments, e.g., where wooden pallets are made, mills purchase cheap pallet logs, i.e., stems of poor form or quality that yield low quality boards.

Ratio Values

Ratios are unique to agroecological economics and this carries into silvicultural agroecology. Chief among these is the land equivalent ratio (LER). The other ratios are economic, but are still comparative.

The advantages of LER and like ratios are that the results are intuitive (where values above one indicate efficiency gains), have simplicity of calculation, and do not require an outside unit, either a monetary value or a discount rate, that can cloud the result. The disadvantage is the lack of data, especially appropriate monoculture yields, that can lessen use.

Revenue

One of many economic ratios is the relative value total (RVT). This is calculated as

$$RVT = (v_a\, Y_{ab} + v_b\, Y_{ba})/v_a\, Y_a \tag{3.1}$$

where v_a and v_b are the values of the annual growth increment for species *a* and *b* respectively. In forestry, this reflects the relative value of larger, clearer, straighter logs. In a biculture, the species (*a*) offering the greatest revenue is the denominator, i.e., $v_a\, Y_a > v_b\, Y_b$.

Costs

The cost structure of forestry plantations is of concern, especially where revenues are in the future and costs weighs heavily upon present decisions. There is no alternative other than to express costs as monetary units.

Evaluation is accomplished through a slightly more complex equation. This is the cost equivalent ratio (CER) where the

$$CER = (C_a/C_{ab})(RVT) \quad (3.2)$$

where C_a and C_{ab} are the costs of monoculture compared with an intercrop of species *a* and *b* (C_{ab}). The other component is the RVT.

A value greater than one indicates that polyculture is more efficient with management inputs while a value less than one has the monoculture being more efficient with the same inputs.

This can be illustrated with applied example of monocultural coffee and with an overstory shade tree (Perfecto et al., 1996). For this, Y_a is the yield of monocultural coffee; Y_{ab} the coffee yields with the trees; C_{ab} the management inputs with trees; and C_a costs without trees. The stumpage value of the tree is not included here. Adding comparative revenues and costs, the calculation of the RVT is

$$RVT = (v_a\, Y_{ab})/(v_a\, Y_a) = \$314/\$1397 = 0.22$$

and

$$CER = (C_a/C_{ab})(RVT) = (\$1740/\$269)(0.22) = 1.42$$

The RVT value is very low (less than 1.0) and would not be supportable as a revenue-generating system. However, the per area cost advantages are shown in the unrefined CER which, adding the appropriate values, gives a ratio of 6.5 (i.e., \$1740/\$ 269). This translates into a CER of 1.42 when revenues (i.e., RVT) are added. Being well above 1.0 threshold, this indicates that the management inputs are used efficiently.

As an applied example, this demonstrates why resource-poor farmers in Central America prefer the low yields and less costly coffee-tree combination. In this case, the positive ecological dynamics generated more than replace costly inputs (e.g., weeding and fertilizer).

When market prices are low, those with a more advantageous cost structure benefit. Silviculturists, with their need to control expenses, are often moved by costs. CER analysis is important in this regard.

Economic orientation

In devising the best silviculture ecosystem, two economic directions are possible: (1) to increase the inputs (costs) to obtain greater output or (2) lessen inputs and reduce outputs. For the first, the increase in outputs and revenue should outpace costs and, for the second, the reduction in costs should be more than any loss in revenue. A workable and attainable ideal is where outputs increase while the costs fall.

Whatever the situation, this movement toward revenue orientation (to increase the inputs to obtain greater output) or toward cost orientation (lessen inputs and reduce output) are recognized silvicultural strategies. The economic orientation ratio (EOR) measures this and is

$$EOR = RVT - (C_a/C_{ab}) \qquad (3.3)$$

As compared with the monoculture, values above zero indicate a shift toward revenue orientation. Values below zero indicate a shift toward cost orientation. This measure does not provide information on the economic effectiveness of the shift. This is obtained using ratios or standard accounting measures for profit and loss.

COMPLEMENTARITY

Complementarity, as measured through LER, is a key standard on how well species cohabit an area and on how plant pairings can be exploited. The technical aspects of complementarity derive from the agrobionomic principles.

Pairings

Early foresters recognized the importance of cross-species tree pairings on complementarity. Schlich (1910) stated that two light-demanding species should not be mixed except on fertile sites and in short rotation. This suggestion may be based on site sustainability and an unhealthy competition for essential resources.

General guidelines aside, strong pairings can transcend site and through mutual gains, overcome a range of site not-to-extreme deficiencies. Weaker forms of complementarity deal with one or a few essential resources. Site selection is more critical when matching or pairing species with weak complementarity. The few guidelines that exist for bicultural pairings are presented in Chapter 9.

The Fundamental Rule of Complementarity

Where two species exhibit a high degree of complementarity, they should be planted in close proximity. Where the complementarity is less, the inter-species planting distance should be greater. This is the fundamental rule of complementarity and this helps determine the optimal planting density.

Generally, high complementarity is exhibited when the presence of one has little effect on a second (i.e., a pairing is capable of an LER greater than 2.0). When this occurs, the planting densities for each should be at a maximum. For example, if the suggested monocultural planting of species *a* is 5000 per ha and, for species *b*, the suggested densities are 4000 per ha, then intercropped these should have 5000 of species *a* and 4000 of species *b* in close contact. Using an easy to illustrate cross-section, i.e., ...*abababa*... where rows of *a* and *b* alternate, the distance between like species is the same as found in a corresponding monoculture.

Where complementarity is less, adjustments should be made in the spatial pattern, planting density, and/or temporal sequence. Although an undeveloped topic, these adjustments are a furtherance on the density aspects of the fundamental rule.

Optimal Interface Distance

For the alternating row (or any other pattern), the question remains as to the best density (arrangement) and the optimal interface distance for the degree of plant-plant complementarity. For this, the LER equation (2.1) can be reformulated as

$$\text{LER} = (Y_{ra}\, r_a / Y_a) + (Y_{rb}\, r_b / Y_b) \qquad (3.4)$$

where Y_{ra} and Y_{rb} are the average yields for rows of species *a* and *b* respectively and r_a and r_b are the number of rows per area for these same species. Y_a and Y_b are monoculture yields for the same number of rows for species *a* and *b* respectively. Figure 3.2 (top) shows the two influences and how they interact (bottom – Figure 3.2) to determine the interface distance that gives the best LER.

Optimal Patterns

Lesser complementarity (LER values between 1.0 and 2.0) suggests alternative planting densities and planting patterns. A sample of cross-sectional spatial patterns that exploit varying degrees of complementarity are:

(a) alternating (one-and-one),
(b) two-by-two, and
(c) atypical.

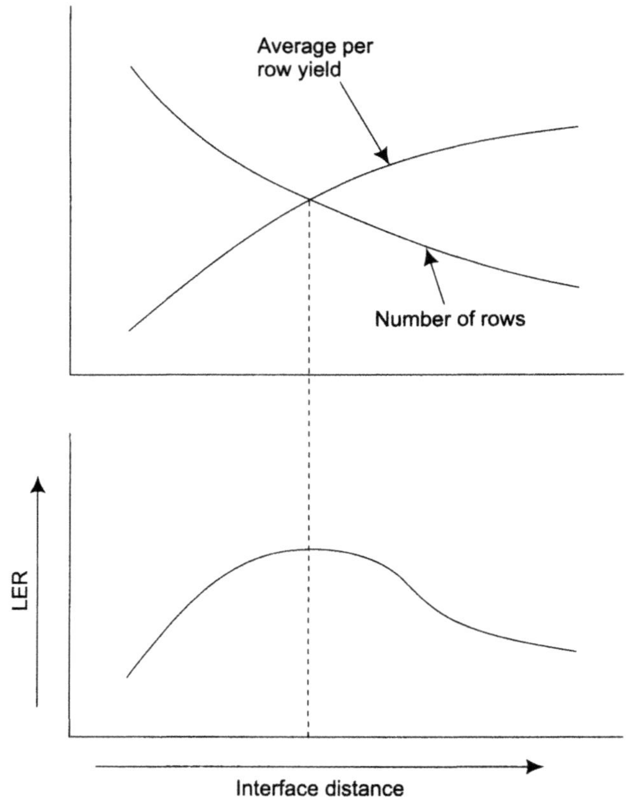

Fig. 3.2 The derivation of LER that compares number of plants per area with growth rate of each species. These convert to an LER function. The optimal density (highest LER) is shown with the vertical line.

Having two individual tree species growing well together, with each at an optimal planting density, is clearly an ideal situation. Where complementarity is less than perfect, modified spacing may be the best option. The easiest to understand option is the alternating ...*ababab*a... arrangement, but with more distance between plants.

A two-by-two arrangement has two species paired as in ...*aabbaabbaa*.... This will not support as high an LER value as an alternate (e.g., ...*abababa*...) arrangement, but may be optimal in situations where height-growth differences are a concern (e.g., one plant shading another) or where some key in-soil resources are in limited supply.

Atypical arrangements are feasible. These have uneven numbers and uneven patterns. The irregular placements, expressed in cross section as *...aaabaaabaaa....* The reason for these are many and varied where the use motives are entrenched in the agrobionomic mechanisms. For example, a ratio of one nitrogen-fixing to one nitrogen-demanding species may be good, but a two-to-one ratio may supply more nitrogen. This suggests an *...aabaabaa...* arrangement when species *a* fixes nitrogen.

Other arrangements accommodate uneven growth, uneven light needs and resource imbalances. The latter occurs if a site is rich in one essential resource and poor in another. The resource profile differences of a pairing may better exploit the site if the species that needs more of the abundant resource is a greater percentage of the overall population.

There can be economic logic to atypical arrangements. If species *a* is three times more valuable than species *b* and, in not wanting to lose or compromise facilitative gains, a planting ratio of *...aaabaaabaaa...* could prove best.

For each pairing, pattern, and arrangement, there is one formulation that gives the better LER. In Figure 3.3, this is demonstrated, comparing two systems (S_1 and S_2) having different density ranges (P_1 and P_2). Each has an optional value and a feasible range. The feasible range is where the LER value is above one. If stipulated, RVT or other measure can replace the LER.

SPATIAL OPTIONS

From the species-based principles presented in Chapter 2, two species coinhabiting a site bring on a host of possible explanations for positive LER values. As discussed in the previous sections, spatial patterns and arrangements can confer non-competitive behavior and an optimized wood-producing stand. There are other facets to this.

Horizontal Patterns

The horizontal patterns are those evident in taking a birds-eye view of an ecosystem. Briefly, horizontal patterns find application

(a) in a monoculture being selectively harvested or as an uneven-aged plantation;

(b) in the planting and harvest of multi-species polycultures, the harvest pattern here usually follows along species lines (there

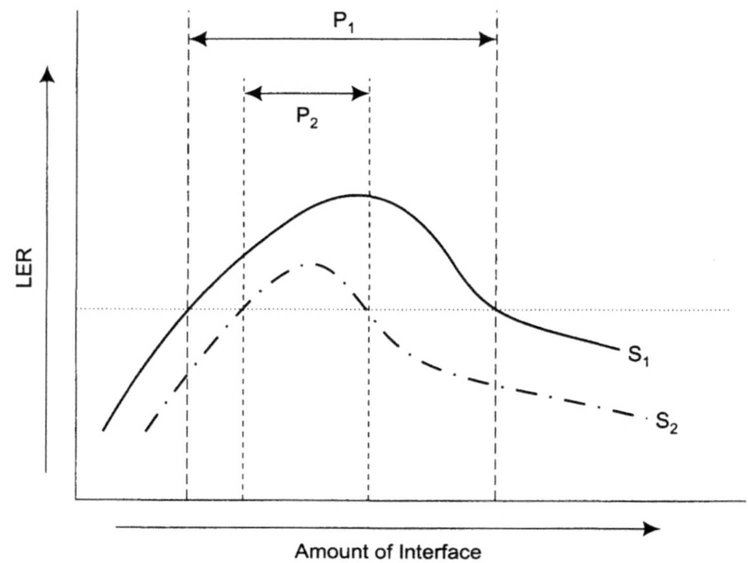

Fig. 3.3 Each system, as defined by the species involved and the spatial pattern and arrangement, has a separate LER function. Two are shown (S_1 and S_2), where the values are above one (P_1 and P_2 respectively) are feasible. Due to a higher LER potential, S_1 is the best choice, all else held equal.

are exceptions, e.g., with a multi-aged, multi-species plantation); and

(c) in natural forests or agroforests where patterns find use in planting and/or harvesting.

These can be roughly divided into two categories, the fine and coarse patterns.

The fundamental rule applied to fine patterns where a high degree of complementarity, a high degree of inter-plant contact and a close interface distance will best capture the positive interactions. The fine patterns are in the right column (Figure 3.4).

Where inter-species complementarity is less, a smaller amount contact and/or interface is better. The coarse patterns, those with a lesser amount of plant-plant interface, are shown in Figure 3.4, right column.

Although the theory is sound and the options known, attaining the full benefit from in-practice patterns is fraught with difficulties. One

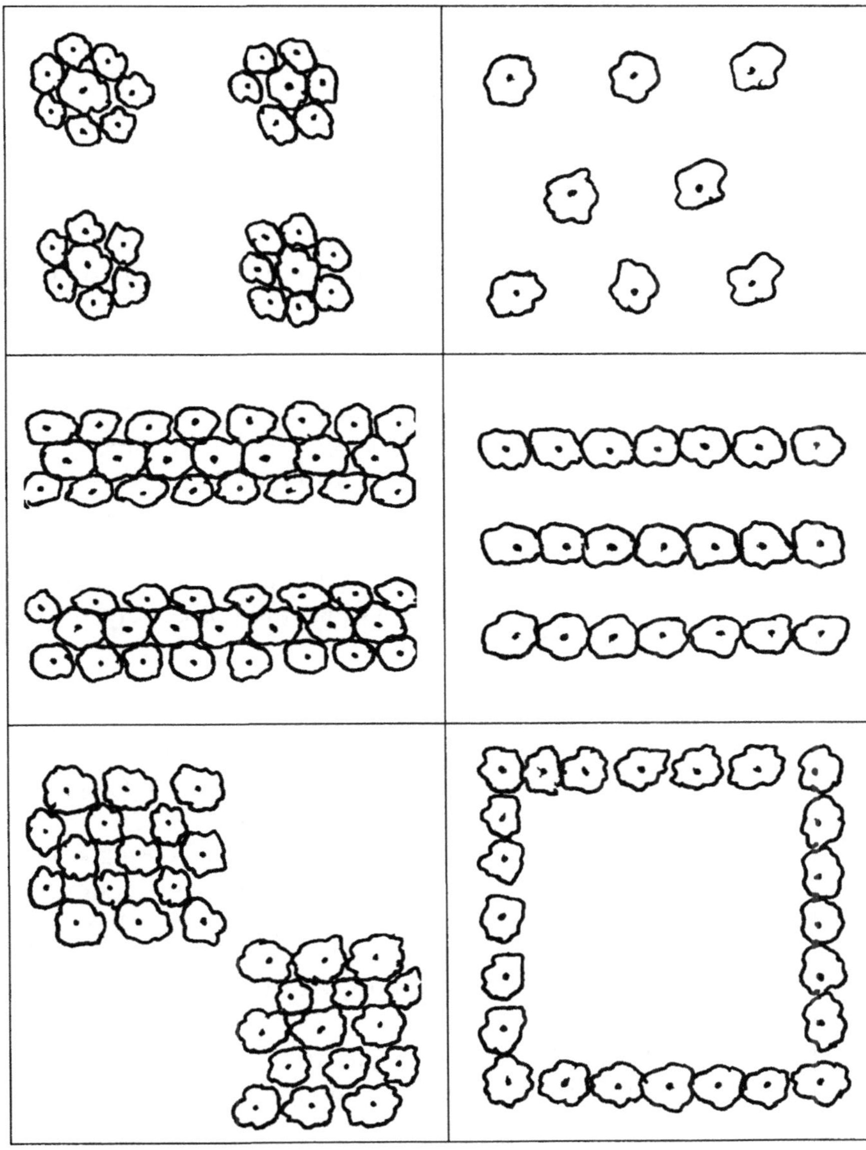

Fig. 3.4 Six basic spatial patterns. These are individual (right top), row (middle right), border (bottom right) cluster or clump (top left). strip (middle left) and block (left bottom) The fine patterns are on the right the coarse patterns are on the left.

being the lack of knowledge on specific tree-tree pairings, those which confer interplant complementarity. Other questions lie in the optimal management of natural ecosystems.

Horizontal Arrangements

For each of the six basic patterns (Figure 3.4), there are any number of pattern arrangements. For example, the row pattern, in cross section, can be *...abababa...*, *...abbabba...*, or *...aabaabaa...* Some possible row arrangements are shown in Figure 3.5. This figure shows two of the *...abbabba...* possibilities (upper left and lower right). These vary as to tree placement and actual dimensions where the dimensions can be inter-row and intra-row.

A three row system, as with *...aaabbbaaabbbaaa...*, exhibits different ecological properties (e.g., in having an edge effect at the inter-species interface). This is categorized as a coarse strip pattern.

For discussion here, only bicultural variations are mentioned. For three-plus polycultures (ecosystems of three of more species), the number of variations grows exponentially as bio-complexity is increased. The spatial options for initiating three-plus polycultures are presented in Chapter 10.

Canopy (Vertical) Patterns

One motive for a spatial pattern is resource partitioning based on comparative height and the light needs of the component species. A starting point is often an equal apportionment of light.

A common situation has one strata above another. These are often of different species where light that diffuses through a uniform canopy (as with open, low light area index canopy) or light that enters through predetermined gaps. For a two-strata system, half of the direct light can be directed to each canopy stratum, for a three-strata situation, one-third can go to each layer.

Within or transcending this context, two underlying vertical patterns exist, these are the midpoint and minimum interface designs. These gain prominence as the number of species increases (i.e., beyond two), with uneven aged systems (monocultures included), or with differences in the canopy structures.

Midpoint

A midpoint design places a shorter species midway between taller species. The pattern is shown in Figure 3.6 (left). Although not well

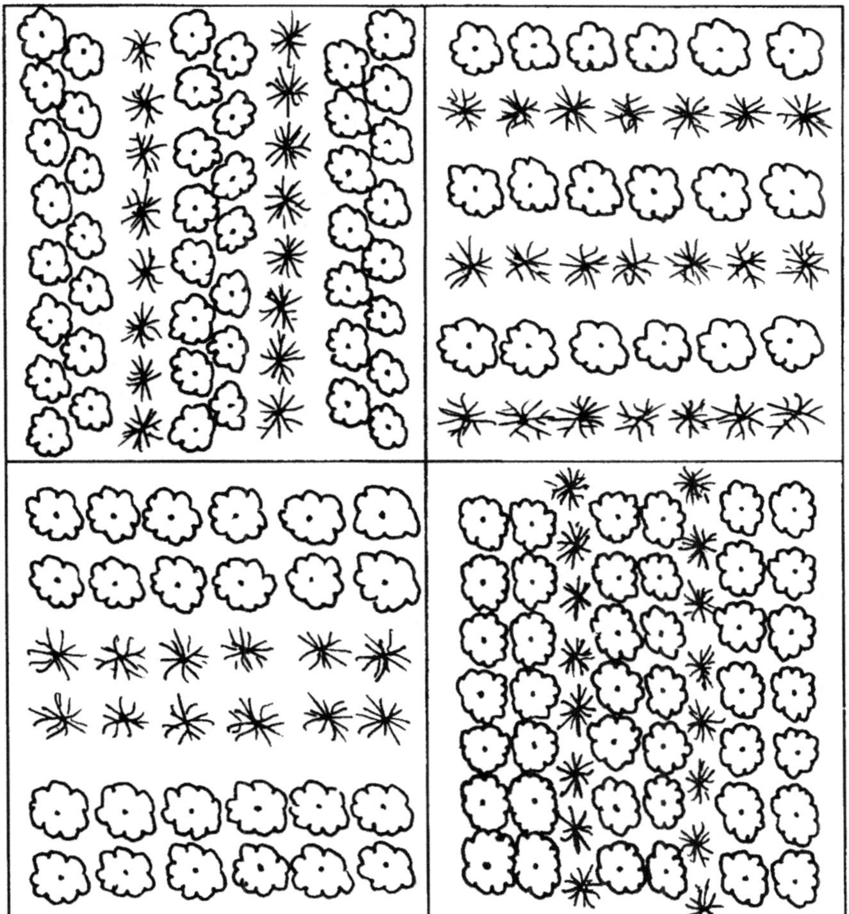

Fig. 3.5 Four possible arrangements of a two-species row pattern. The variations include planting densities (intra and inter row) as well as the cross-sectional ordering.

studied, this pattern seems best where light is to be evenly apportioned among component species and canopy spread not a concern.

One application has a species (species *b*) that gathers horizontal light located above one that prefers vertical light (species *a*). This may favor a greater distance between the upper layer trees, according a system cross section such as ...*aaabaaabaaa*....

Minimum interface

A minimum interface design has a shorter species immediately adjacent to a taller plant. In pure ecosystem form, the arrangement of species is such that plants progress downward; the tallest, adjacent to the ensuing tallest, next to a shorter species, or on downward. This pattern is illustrated in Figure 3.6 (right). Carrying this forward, Figure 8.4 shows a minimum interface within an uneven-aged monoculture.

A minimum interface has some clear applications. Where many different species are close together, the effect is one of suppression for the taller species, denying full access to below ground niches. This effectively transfers a portion of the essential resources to surrounding, shorter plants. Therefore, application may be best with more biodiverse systems, those with an active understory rather than in all-tree, like-aged ecosystems.

Other applications are more in tune with silvicultural needs. A resource compatible guide species, selected and positioned to improve the stem form of the inner tree, can hem the main species. A third, shorter species may augment this effect. Planting distance is a function of complementarity where the more resource competitive species are wider spaced, the more compatible species are closer.

Stem-hugging, stem-improving guide species can surround high-value trees. Most tree species will benefit from reduced side branching. This service is far less needed with species that predominantly gather vertical light and have a naturally clear stem.

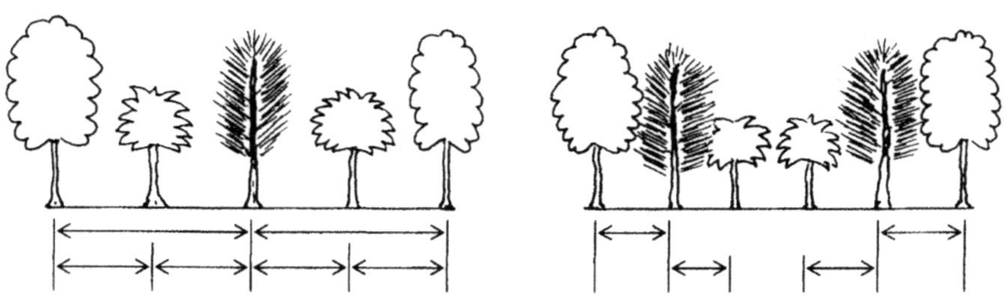

Fig. 3.6 The two basic canopy patterns. These apply only to systems of three or more species or to uneven-aged systems with three or more age classes. On the left is the midpoint design, on the right is the minimum interface design.

DISARRAY

Outside the natural forest, agroecosystems in some regions exhibit spatial disarray. Besides being a cultural manifestation, this takes advantage of ecological dynamics to reduce risk and increase production when resources are variable. In some aspects, disarray is the spatial equivalent of genetic diversity.

Reasons

With ordered systems, the amount of resources available to each plant is determined by density and, through this, the optimal planting spacing of the component species is determined. For any tree monoculture, the optimal density is set based upon average years and average resource availability. If available resources are not close to this predetermined seasonal mean (intra and/or inter-seasonally), the chosen spacing is not optimal, nor is growth.

With disarray, uneven spacing offers more essential resources to some plants, less to others. When resources are readily available, some trees partake in abundance, some have less. When resources are restricted, those with wider spacing still produce well, those crowded may experience good (with facilitation) or poor (without facilitation) growth. The net effect is better average growth in those years or seasons that are not resource average. This is shown in Figure 3.7, where more growth occurs in resource scarce or resource abundant situations, less in average years.

With more planned and unplanned biodiversity, the effect will be magnified. Where resource inputs are relatively constant, order or semi-disarray may be the better alternative. Where average years are seldom encountered and there is more climatic and soil variation, full disarray may serve best. There are other aspects to this, e.g., disarray may help thwart herbivore insects and plant diseases (see Chapter 7).

Types

In full disarray, systems lack any discernible uniformity (Figure 3.8, lower left). Lesser degrees of disarray do exist. For example, partial disarray can have a uniformly planted primary species while the secondary species (one or more) exhibits no pattern (Figure 3.8, upper right). Another variation (not illustrated) exists where spacing is not exact, but an underlying pattern is discernible. A final, more common,

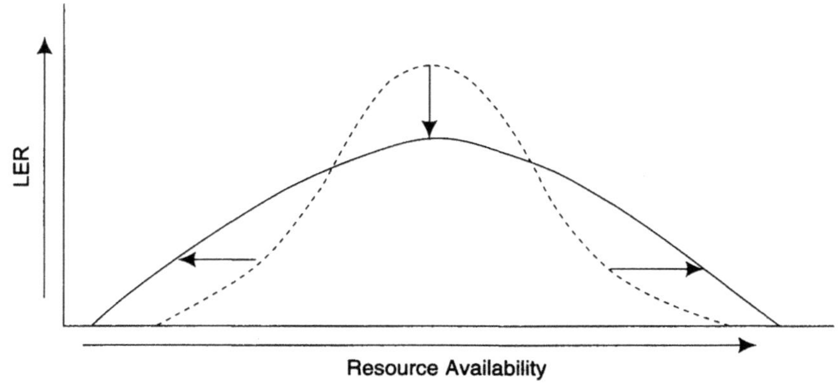

Fig. 3.7 The theoretical gains from disarray. The inner growth function (dotted) is for ordered systems, the outer for disarray. When a single essential resource is abundant or scarce, better growth can be expected with disarray. Under normal conditions, ordered systems tend to yield better.

form has disarray coupled with increased density (Figure 3.8, lower right).

Disarray is found in some regions, but motives may vary. With polycultures of three or more species, the landuser may have no idea as to the optimal spacing or pattern. Instead, disarray coupled with increased density may serve as a surrogate for what is not known.

The only thing that can be said with certainty is that, besides numerous unstudied in-field examples, almost nothing outside the basic theory is understood. Implementation comes with caveats and, although theory suggests major gains in prescribed situations (e.g., large variations in seasonal rainfall), practice may not be so kind.

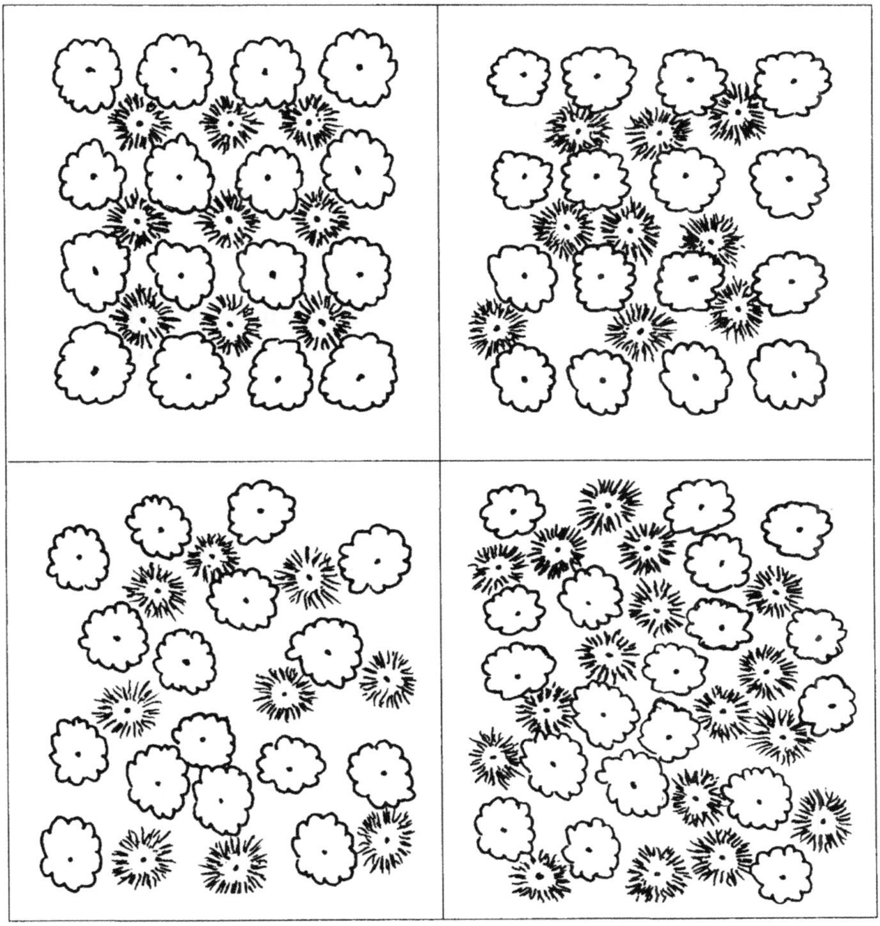

Fig. 3.8 Shown are an ordered biculture (upper left), an ordered upperstory with disarrayed understory (upper right), complete disarray (lower left), and complete disarray with increased planting density (lower right).

CHAPTER 4

Temporal Dynamics

In the previous two chapters, the underpinnings of non-temporal dynamics are reviewed. Due to the long-term nature of most wood-producing systems, temporal dynamics can be more of a force. Much of what is utilized in-field has parallels with the successional dynamics of what occurs within nature. Practitioners can and do use these dynamics for positive gain.

REALIZED NICHES

Bicultural systems have realized niches, but in the narrowest sense, i.e., looking only at competition between two dissimilar plants. In complex ecosystems, niches are realized in a variety of ways, some promoting positive inter-species dynamics, others are not so giving.

These relationships also apply across time. Depending upon the planting sequence (e.g., overlapping or sequential) niches change as an ecosystem progresses and, as with spatial niches, there is opportunity for silvicultural exploitation.

NATURAL SUCCESSIONS

Trees are long lived and must adapt to non-temporal ecological dynamics in their chosen ecosystem as well as those of the temporal plane. One implication is with evolutionary-derived ecosystem membership where membership is also a temporal affair. Those species that are niche comfortable, i.e., have slight ascendency, in an early successional phase may not be at home over time. However, early-stage species may set the site requirements that favor or provide ascendency for latter-stage species. That is, the residual vegetation after a disturbance may push an ecosystem in one direction or another (Platt and Connell, 2003; Svenning et al., 2004).

Oak following pine and pine following oak has been observed for a long time (Emerson, 1846). Peterson and Squiers (1995) found that, when white pine followed poplar as the next successional step, the white pine grew better. Callaway (1992) and Gómez-Aparicio et al., (2004) showed that mid-successional trees did better, presumably resulting from a changed ecosystem, with the facilitative affects of pre-existing shrubs.

A positive effect may be temporally one-on-one (where one follows the other to achieve the desired result) or ecosystem related (where one combination of plants follows another, e.g. when trees follow shrubs). A corollary of this is that, once cut, some climax ecosystems require a re-establishment sequence before they can again dominate. All the same, the course taken may vary.

Phases

The growth progression of a natural forest is one of overlapping successions of different species. After a disturbance, the New Zealand species kauri must wait a considerable period and for a sequence of events to take place before again becoming dominant (Ogden and Stewart, 1995). The first phase takes about 50 years and contains mainly *Leptospermum scoparium* and *Kunzea ericoides*. The second phase, lasting about 150 years, has kunzea as the primary species with a kauri understory. This may serve as a facilitative species that promotes the growth of kauri. The climax forest with dominant kauri reaches maturity after about 350 years.

What has been designated above are standard forest successional phases or stages. These are:

(1) scrub phase,

(2) early successional (pioneer),

(3) mid-successional, and

(4) the final climax phase.

As with all natural systems, exceptions are the rule and a four-stage succession may not be universal. Some ecosystems may have a more abbreviated overall schedule, go through one or more of the individual phases a bit quicker, or have fewer (or even more) phases.

Scrub phase

After a major disturbance, a host of non-woody plants and woody shrubs can invade an area. These can be annual or perennial and may suppress tree growth for a few seasons or for a number of years.

Although trees are not present, the ecological role that herbaceous plants play is useful in silviculture. Some actively or passively suppress other vegetation and, with this attribute, potential cover crops are identified.

Early successional

The scrub phase eventually succumbs to an early succession of pioneer trees. Usually fast growing, light demanding species, that, once established, compete well against herbaceous plants. These may colonize areas in fairly pure stands.

The early successional phase can be visually determined, usually by the residual non-wood plants and shrubs interspersed amongst taller pioneer species. This succession may also be open with patches of the scrub vegetation and clumps of early successional trees. Color Plate 4.1 (a) shows early successional temperate forest with a mix of shrubs and pioneer trees (the white birch being a pioneer species).

Mid-successional

The pioneer species give way to another generation of trees. These have the ability to survive and grow with less light and can thrive as an understory to the pioneer species. This phase is defined by residual pioneer species along with an understory and partial overstory of hardier, slower growing, shade tolerant trees. Color Plate 4.1 (b) shows mid-successional forest where soon-to-be climax hardwoods are succeeding in a pine stand.

Climax phase

A natural forest ecosystem eventually reached the point where species mix, barring disturbance, remains more or less static into the future. This is the climax forest.

The climax phase can be defined by one or a mix of specific dominant species where the ecosystem excludes other than the one or more dominant species (Baker, 1934). In tropical humid forests, with their proliferation of species, this is harder to gauge as some light-demanding, earlier successional species may persist in the understory, waiting for a gap to occur (Dalling et al., 2001). The common visual measure in untouched tropical high forests is the presence of very large diameter, tall straight trees and the existence of an emergent layer (those few trees that tower above the upper, continuous canopy).

Temporal Divergence

As mentioned, the stages outlined here are far from universal, variations can and do exist. Some regions may only have single succession where a forest proceeds from pioneer species through to the climax species in a single bound.

Another variation is in the ordering of the phases. Some species are associated with a particular phase, for others, this is relative. What is a dominant species in one area can be successional in another. Baker (1934) mentions that, in the Pacific region of the USA, redwood is dominant against Douglas-fir within the redwood range. Outside this range, Douglas-fir and lodgepole pine are climax species. In another example, pines can be a step in a succession or a climax species. In a study of vegetative change in the Mediterranean region, Moreno et al. (1993) were surprised when pines succeeded species that were thought climax.

Differences in succession in what may seem to be attribute-like sites can be brought about by less than obvious influences. Soil differences, tiny climatic disparities, and the introduction of an outside species can all contribute.

Dictated vs. Event-Driven Successions

Depending upon the specific area, vegetative succession can be viewed either (a) a set progression or (b) a series of directional options. These may be the result of a long evolutionary development where, if climatic variation and traumatic events were common, multiple phase directions were naturally put into place, if not, only one track originated. This may be another element in ecosystem membership.

With the mainstream or one-track hypothesis, there can be major vegetative differences in a earlier phase, but the next few converge back to the sequential norm. In an Australian succession, Grant and Loneragan (2001) found some evidence to support this hypothesis. In this case, pre-burn vegetation foretold the composition of post-burn regeneration. In other words, fire did not alter the vegetative future.

Other temporal systems may be omni-directional, relying upon unfolding events to set the plant content of each successional phase.

Usually traumatic and/or climatic events cause an ecosystem to unfold in different directions (Strout et al., 1975).

Dyrness (1973) found differences in vegetative establishment if a site was (a) soil undisturbed, (b) followed logging, or (c) followed logging and slash burning. In these multi-directional cases, much depends on the intensity and type of disturbance, seed sources, and the ability of seeds to survive and spread.

As a result of differing traumatic events, the intermediate and end points may greatly vary between like sites. Fire has turned the successional order from yellow poplar to oak (Brose et al., 1999) or where successional pathways changed in one of four directions (plant community types) in response to fire frequency (Frelich and Reich, 1995).

TEMPORAL THEORY

From the successional phases, with variations, a modest theoretical base emerges which explain the dynamics of temporal silviculture. This stems from a number of components and interactions along the successional timeline.

Rotations

Rotations are visible in even-aged monocultures, far less so in complex, unevenaged polycultures. For plantations, these go from one bare ground phase to another. In managed natural forests, this is time taken for all species to be replaced (cut, regenerated and cut). This is visible with clearcuts. With selective harvest systems, the full rotation is less conspicuous.

In plantations and clearcuts, the cutting cycle equates with the rotation, i.e., each cutting cycle is also a rotation. In selective harvests, a rotation can be composed of numerous cycles. For example, if it takes three cutting cycles to remove all the trees that existed at the start of the cut-regeneration sequence, this will be a single rotation (as in Figure 12.1).

Natural Equivalencies

In commercial forestry, the scrub phase is avoided if at all possible, with the noteable exception of simple taungya where non-woody invasive weeds are purposely supplanted by crops. Many of the common plantation species are pioneer trees that grow fast from open sites. The resulting plantation ecosystem, provided that enough biodiversity is present, may be ecological equivalent of the first successional phase.

Bicultures of primary species duplicate, either successionally or with set phase, that which occurs in overlap between two successions. This can be pioneer or mid-successional. In all cases, there is a clear link with successional phases and with silvicultural treatments. This is shown below.

Timeline standing	**Silvicultural variation**
Pioneer phase	Single rotation (monocultural)
Pioneer phase	Single rotation (bicultural)
Latter pioneer	Single rotation (bicultural)
Pioneer to mid-successional	Series rotation
Mid-successional to climax	Series rotation
Pioneer to mid-successional	Overlapping series
Mid-successional to climax	Overlapping series
Pioneer to climax	Overlapping series
Scrub to pioneer	Taungya (simple)
Climax	Continual (closed canopy)
Scrub to climax	Continual (patch)

SILVICULTURAL TEMPORAL VARIATIONS

It is better to work with, rather than against, natural ecological forces and the temporal variations strive to do this. In discarding basic ecological principles, this task is made more difficult. To keep within natural guidelines, the following temporal strategies are species and objective accommodating.

Single or Discrete Rotations

With single or discrete cropping, an ecosystem exists for a set period of time with no planning on future land use or where the subsequent rotations are a continuation of the same ecosystem type. Figure 4.1 illustrates this for two rotations, here each temporal segment of a shrub-facilitated biculture ecologically duplicates that done in the previous rotation.

The lack of a planned sequence eschews some of the advantages that planned and ecologically directed rotations provide (e.g., in

Fig. 4.1 A temporal sequence (top to bottom) for single or discrete rotations. This has two like plantations of shrub-facilitated bicultures, one following another. With discrete rotations, there is no presumption of inter-period ecological gains.

temporal biodiversity). This applied to the control of insects and diseases and in maintaining soil fertility.

As an ecological force, tree removal duplicates natural destruction where ecosystems return to a near bare ground state. The key is in skipping the non-forestry scrub stage.

Many, but not all, latter successional species are suited to a more exposed plantation life. As previously mentioned, those that are employed are often the fast growing, early successional (pioneer) species that are well adapted to growth after vegetative destruction. This can include complex multi-species ecosystems; mixes of non-competing pioneer species or, as in Figure 4.1, some facilitative-productive mix.

Series Rotations

The idea that one species naturally follows another can be harnessed for productive good. In contrast to a natural ordering, a series rotation has a bare ground juncture between each planting. Figure 4.2 shows

Fig. 4.2 A plantation series where a shrub-facilitated system sets the ecological stage for a monoculture of a latter successional tree-species. This presupposes ecological gains from the pre-planned sequence.

series rotation where one system (a biculture) is followed by a second (a monoculture) in a planned progression.

With the longer rotations of forestry, the ecological shortcomings are less than those of seasonal agriculture. Still, the mechanisms of advantage can find successful application. This occurs when one rotation sets the soil or site stage for another. For example, badly degraded or eroded soils can be made suitable for a high value planting once the site has been occupied by a soil-improving species (as in Photo 6.1).

Other uses are less an immediate environmental remedy and more a plan to bring biodiversity to a monocultural landscape, as with a series of monocultures of different species. Except for a brief period of bare ground (and an associated erosion danger), a series plantation approach, one following naturally-set successional patterns, can be more in natural harmony than a progression of like monocultures.

Overlapping Series

Viewed through a successional lens, the rotational possibilities bring to the fore a series of overlapping patterns, each contingent on previous land use, each a species component of an established successional phase. Among the alternatives are:

(a) a series of monocultures with a brief overlap (as shown in Figure 4.3),

(b) bicultures with more pronounced overlaps and a continual vegetative cover,

(c) various agroforestry possibilities that mix treecrops (e.g., fruit trees) with forest trees, or

(d) a natural forest series where one intense silvicultural treatment leads into another with a smooth transition.

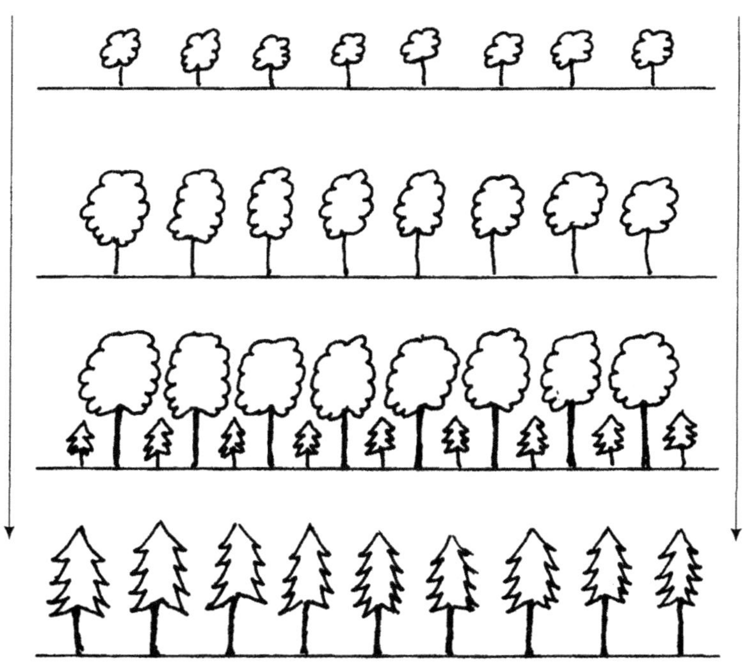

Fig. 4.3 A overlapping sequence where, instead of a bare ground period, one system temporally merges with another. In this sequence, the climax or ending species is planted a few years before the initiating species is cut. The reliance on inter-period facilitative gains means that the species employed are usually naturally successional.

In addition to ecological and yield gains (e.g., less erosion danger in the temporal juncture or higher, temporally efficient LER value), an overlap can shorten the overall rotational period. The cues for this come from nature where, as in Photo 4.1, one species is naturally succeeded by another.

Taungyas

Within the interface between forestry and agriculture, there is plenty of scope for temporal variation. The replacement of scrub herbaceous plants with herbaceous crops is not a large step. Many early farmers and foresters found taungyas an attractive option well before the term was coined (in the mid-1800's). The possibilities go beyond the scrub phase association. Chapter 11 delves into these, Figures 11.1 through 11.4 shown some of the temporal progressions of the taungya.

Continual

A characteristic of most natural climax ecosystems is in having no clear beginning or end. Without a major, unforeseen, traumatic event and with an appropriate silviculture treatment, it is possible to maintain a climax forest into the foreseeable future. This can be a management strategy where protection ensures that traumatic events, except logging, do not intercede and treatments are light enough so the ecological character of the system remains stable.

The other option is, through management, to keep an early successional phase in place. Keeping this continual requires a more intense effort and heavier touch. In an event-driven ecosystem, the right combination of induced events can accomplish this in perpetuity (e.g., low intensity fire with natural reseeding). Lacking this successional handle, the ecological character may be more traumatic and less continual (as with manual replanting).

Mixed systems are possible and continual. There is the option of keeping the forest as a series of small clearcuts (i.e., a patched forest), each beginning the process of successional development. These areas may be harvested at one or more successional points.

Although more akin to climax situation, monocultural plantation of a climax species can be maintained for an extended period. Figure 4.4 shows a continual monocultural plantation where harvests are maintained over time through continued selective cutting. A multi-species version of Figure 4.4 is entirely possible.

Photo 4.1 A pine stand with a natural occurring oak understory. Natural pairings are indicative of those species that grow well as successional, overlapping partners.

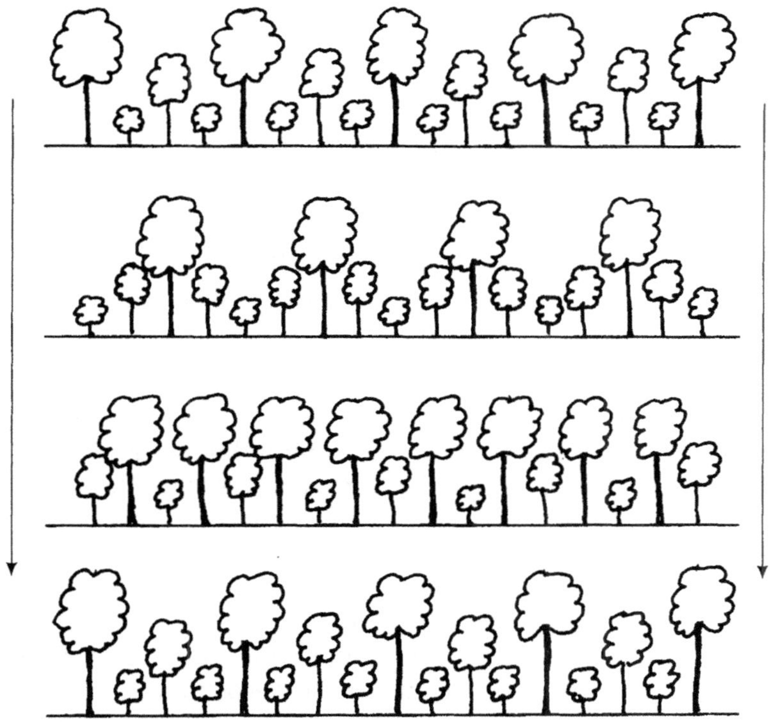

Fig. 4.4 A continuous system where there are numerous cutting and planting cycles within a very long or non-ending rotation. Each step, four are shown, involves cutting and replacing the tallest trees.

SUCCESSIONAL (TEMPORAL) ECONOMICS

Single rotation monocultures are favored, in part, because of their commercial simplicity. Attractive to banks and other financial institutions (McNeely, 1993), commercial simplicity does not always exploit ecological advantage.

The economics of succession can be difficult. The lack of any meaningful analysis for the most complex of systems (i.e., the agroforests) may limit commercial interest.

For evaluation, there is a long recognized need to take into account time value. Basically, this is where a present payment does not often equate with one due in the far future. This is the time value of money.

In silviculture, time value is part of the decision process where costly current activities yield future income. Time value is also used in determining how revenues and costs, set unevenly upon a time line, can be reconciled and compared.

Time Value

The need to reconcile time differences and uneven revenue and cost patterns is accomplished using net present value (NPV). The basic formula is

$$\begin{aligned} NPV &= -C_0 + (R_1 - C_1)/(1 + r) + (R_2 - C_2)/(1 + r)^2 \\ &\quad + (R_3 - C_3)/(1 + r)^3 + \cdots + (R_n - C_n)/(1 + r)^n \end{aligned} \tag{4.1}$$

where revenue is the different time periods (year 1 through year n) as R_1 through R_n and where the costs, starting in present (i.e., C_0), proceed through year n (C_n). The mechanism of relative value for the time periods is the discount rate (r) where, the higher the rate, the greater the emphasis is on the earlier years and the lower the comparative value of future periods. Rates are percentages in decimal form (e.g., the common value of 4% = 0.04).

Time value is a staple among bankers and finds informal use among the general population. Without formal analysis, most landusers understand that long timeframes are barriers to the utilization of some tree species and forestry ecosystems.

Despite widespread cognizance, the technique should be based on practical considerations. If a plantation has a projected 75-year rotation, there is little inducement among landowners, especially those over 50 years of age, to make this investment. This is true even at very high NPV-determined rates of return.

Governments may want to make a long term investment, not only for the revenue, but for subsidiary benefits (e.g., as part of an environmental effort). Other situations, those where this investment may be favorably viewed, have the 75-year rotation underway and industry and/or community needs are better served if the rotational series (and the species being utilized) does not lapse. Some European forestry efforts, long underway, fall into this category and profitably defy NPV analysis.

Spatial and Temporal Complementarity

Plant-plant complementarity is usually defined within set period with both species present. There can be two forms of complementarity, that of a given moment and that which crosses periods.

There are some general guidelines that relate intra-period complementarity to the inter-period version. If two species are complementary, growing well together, they will generally grow well as temporal neighbors although the temporal ordering may be important. This can occur when resources are derived from different sources (e.g., a nitrogen fixing plant paired with a nitrogen demander or when plants have roots in different zones).

Examples abound where nitrogen fixing species are paired with one seeking this element. This can be spatially, temporally, or some combination, e.g., a staggered planting where a certain level of nitrogen fixation occurs before the second, the demander, is in place. This could be the case in Figure 4.2 where the shrub in the shrub-tree biculture is especially good at capturing nitrogen.

Inter-time complementarity also happens when one species uses a lot of one resources (resource *a*) little of another (resource *b*) and the second species requires little of one (resource *a*) and more of the other (resource *b*). These mechanisms function across time periods. Figure 2.3 can be temporally reformulated to illustrate this situation.

Some species pairings, those that do not interrelate well spatially, can work well temporally. Where water or light can be the limiting resources that foretells failure. In the temporal plane, light and water may be less of a factor in temporally separate monocultures or polycultures, bringing to the fore positive complementarity in mineral resources (as with the resources *a* and *b* example, paragraph above).

If needed, the LER equation can be reformulated to measure temporal effects. This is as

$$\text{LER} = (Y_a/Y_a) + (Y_{ba}/Y_b) \tag{4.2}$$

where Y_a and Y_b are seasonal growth from monoculture plots of species *a* and *b*. The yield of species *b* following a temporal exposure to species *a* is Y_{ba}. This measure can be applied to any sequential planting, including discrete rotations. The caveat being that the rotational sequences being compared are of similar duration.

Growth and Density

As an ecosystem ages, trees grow and, although the spatial pattern may remain optimal, spacing may not. Less than optimal planting densities can encourage side branching while overly high densities can result in many spindly, rather than thicker, higher valued stems. Most often, there is not one optimal density, but where the optimal density starts at some high number and declines over time. Photo 4.2 a and b show sub-optimal densities: one with too wide a spacing, the other too dense.

In the non-static version, optimal density takes on an aspect of compromise, where (a) each plant has access to sufficient essential resources to keep a high rate of growth over time while (b) essential resources are not wasted (e.g., light not striking tree leaves should be kept to a minimum) to maintain the highest possible per area productivity. Also, there are (c) stem quality issues where some crowding eliminates side branches, yielding straighter, knot-free, and more valuable logs and (d) cost issues where it is more cost effective to harvest in less crowded forests or plantations.

The process is usually subdivided into (1) pre-commercial and (2) commercial. For the latter, there is a strong impetus to undertake this economically positive activity. Unfortunately, this occurs mostly in older stands.

The pre-commercial thinning of small diameter trees is more problematic. Various means exist, ecological and otherwise, to promote a more advantageous spacing and to circumvent costly and non-profitable thinnings. A lot of this relies upon biodiversity. Discussion of nature-accomplished management needs continues in Chapter 6.

(See caption-next page)

Photo 4.2 Two non-optimal planting densities: (a) the trees are too far apart as evident through excessive side branching; (b) an overly dense spacing that results in thin, spindly stems, and poor per tree growth. Photo 4.1 is of a better spaced plantation ecosystem.

CHAPTER 5

Use Concepts

Going beyond the agrobionomic principles, species and ecosystem governance, and the spatial and temporal theories, a host of other concepts underpin and advance silviculture. These are yet another step in the transition from pure theory to useful field practice.

As in Chapter 2, ecosystems can be roughly grouped as species or ecosystem governed. In this chapter, the concept takes another form, with less concern regarding ecological governance, directed more toward explaining economic roles and avenues of practical implementation.

SPECIES VIEWPOINTS

Under species governance, the agrobionomic principles describe the relationships between individual plant species and how these are managed around one-on-one, plant-on-plant interactions. The species view also underwrites how trees are selected and paired, however, not in agroecological terms. The view is one from both a practical operational and a market perspective.

Primary and Secondary Species

The breakdown of an agroecosystem into primary and secondary species is a trait of agroecology. In most cases, the difference is unmistakable. The primary species is of higher value, more desirable, and is given, over the full timeframe, greater access to essential resources. In contrast, the secondary species is of lower value and, in an ecosystem, has a more subservient ecological and economic role.

Although this relationship is clear within most ecosystems, there are complicating factors. These are when

(1) the primary-secondary roles reverse over time as when,
   (a) in a simple switch, the ecological functions are transposed; or
   (b) the secondary species becomes the primary species upon removal of the original primary species;
(2) the relative roles are more related to age or size than species type.

Species Selection

Both primary and secondary species can be chosen for their wood outputs. Equally likely, a secondary species may be singled out based upon facilitative services to (a) the primary species (species governance) or (b) the ecosystem as a whole (ecosystem governance). These may also be selected for (c) some non-woody product. Treecrops (fruit and nut trees, rubber trees, bamboos, palms, etc.) are prominent under the latter heading.

Wood properties

For any one end use, some woods serve best. The ideal properties for furniture woods are light weight, decorative (refers to the surface color and grain pattern), fine textured, machines and finishes well, and has dimensional stability (minimal shrinking and swelling). The most expensive furniture is manufactured from walnut, cherry, true mahogany, and other species with close to ideal properties.

For wood flooring, color and finish are among the desired properties, as is hardness (to resist denting). Oak, having all these properties, is the standard for flooring in many countries.

The notion of choosing the right wood for any given end use extends to those with specialized, hard-to-duplicable attributes. For example, heavy, dense black locust is best employed when wear is a problem (e.g., stair treads) while impact-resistant ash finds use in hammer handles and some sports equipment (baseball or cricket bats). In a perfect world, instead of a few general-purpose species, woods would be well matched, through properties, with their best end use.

In theory, wood properties should relate to worth. In practice, perceptions, more than wood properties, often drive local markets and these are often regional. In central Europe, beech is a prized wood, oak lags in value while, in North America, the opposite occurs.

Since the primary silvicultural species has the higher market value, beech would be a leading primary species in European silviculture. When grown together with oak, the comparatively higher worth of beech trumps oak, beech being the primary species. In the absence of beech or some other more value species, oak assumes this role.

Woods with unique and useful properties, but in low demand, are routinely relegated as secondary species. Grown in conjunction with a higher value species, these still up the worth of the overall ecosystem (i.e., the RVT).

Field properties

In the real world, wood properties and demand are only one facet in silvicultural selection. Practitioners most certainly prefer trees with a high end-value, but more importantly, they require good growing properties. Uncertainty regarding future markets and the increased emphasis on engineered wood products can mute the requirement for high-valued species.

Temporal compromises also cloud the value-selection relationship. To produce an equivalent volume of wood, continuing rotations of fast growing trees require less land area than slow growing woods. Under this premise, long rotations (e.g., 100 plus years) are tolerated only where nature sanctions little else and where cultural values are conducive to long timeframes.

Growth time, rather than wood quality and/or market strength, often control species selection. The number of pine plantations worldwide is indicative that wood properties are less of a selection criterion. Pinewood is not exceptionally unique nor does it have the auspicious property uses of other woods. It dominates as a plantation species because (a) the tree grows fast, (b) a lot is known on propagation, (c) tried-and-true varieties exist, (d) the wood has any number of end-use applications, and (e) it has an assured and steady market value.

A more dramatic example is eucalyptus. This is a tree with poor drying properties which curtails many end uses (e.g., as lumber). This is a case where growth properties, in specific good form, rapid growth, and an ability to grow well on very arid sites, favors large scale plantings. This also supports the adage that, if enough is available, the market will find applications. Eucalyptus, in fiber form, is utilized for paper and in engineered particleboards and, in solid form, for poles.

For a secondary species, growth properties are important as long as the species has a lesser market value. Once this requirement is met, the second species (one or more) should be complementary with the primary tree. Slow growth can be tolerated or even be an advantage in an overlapping plantation sequence.

Species, such as walnut and oak, could fit this role although, if the market value is high, these tend to shed the secondary species designation. If this occurs, this may be indicative of a role reversal, i.e., a transition from a secondary to a primary species once a faster growing component is removed. This comes about in pine plantations with oak understory where the slow growing, high-value oak has, in the initial phase, a secondary designation.

As a deciding factor in what and how long to plant, log size should also be mentioned. In some regions, the greater worth lies with pole-sized stems. Size limits exist even when sawmills are the main customer. Large logs are cheaper, per unit of wood, to process but, when the milling equipment is designed for certain lengths and diameters, log dimensions can be more important than wood type.

Facilitative trees

Discussion to this point assumes the secondary tree has market value. This may not always be the case. In multi-species applications, a secondary species may be included less for wood output and more for facilitative gain. Increased primary species productivity, improved stem form, the elimination of unprofitable thinnings (for an illustration, see Figure 6.1), and/or early phase weed control may be enough economic justification for the inclusion of a purely facilitative species addition.

Treecrops

Trees that produce other products of value expand upon the silvicultural possibilities. As with previously mentioned rubber plantations, the main product is latex but, because of the quantity available, the wood is in wide use. Another treecrop are various palms (coconut and oil are commonplace examples). The wood can be a commercial product where plantations are common but, because of quality issues, palmwood is found only in local markets.

Treecrops encompass a wide range of species; the aforementioned rubber and palm trees along with fruit and nut trees of all types.

In silvicultural situations, a secondary species can be added both for its treecrop (i.e., non-wood output) value and for its facilitative input (e.g., being of short height, these serve admirably as guide trees).

In an interesting aside, this concept has direct implications in cultural agroecology and community forestry. (Silva et al., 1985). In exchange for the fruit or other non-woody outputs from the secondary treecrop species, locals provide free management services for the plantation (e.g., weeding, thinning, and pruning). This reduces yearly management costs, the wood revenue, and what should be an enlarged profit from the closing wood harvest, goes to the owner.

In a type of role reversal, a forestry species, one without any product value other than wood, can serve as a secondary species in treecrop setting. A slow-growing forest tree amongst palm or rubber trees epitomize the concept. For this, both are harvested for wood at the end of a 30- to 40-year rotation. Rarely encountered, these remain a possibility.

Multi-purpose trees

The multi-purpose species has the tree as a wood source, but also doing something else (Huxley, 2001). The additional roles can be some ecological function (e.g., a habitat for bird species), agroecological function (e.g., a farm windbreak), a social function (e.g., a village shade tree), a religious function (e.g., part of a sacred grove), or have an economic purpose (e.g., fruit yields). The latter also extends the bounds of traditional silviculture through the ordering of the primary and secondary products.

Supplementary additions

Some non-woody products can be obtained without impinging upon primary and secondary species or affecting an ecosystem in any major way. Bees and honey production is a classic example. Some tree species, e.g., eucalyptus, are known for their honey quality. Another supplementary output, rattan vines, used to manufacture cane products, are grown in the canopies of established tree plantations (Mohd Ali and Raja Bariza, 2001). Although outside the primary or secondary designation, these additions may still push users toward tree species or silvicultural practices with supplementary product potential.

Desirable Plant Characteristics

Species selection usually depends on a combination of factors. Wood properties, high future value, growth rate, and a knowledge of the wood and growth properties all contribute. In addition, certain trees tend to be fashionable, i.e., prevailing in recent literature, and this influences the selection in tropical regions. Past and present examples are teak, gmelina and leucaena; each of which found favor at a point in time.

Fashionable trends aside, selection can be based on desirable plant characteristics (DPCs). This is less judgemental, more quantitative, and is especially potent when there are many candidate primary or secondary species.

The idea is to best fit end-use requirements (sampled below) with the growth attributes and wood properties (the DPCs) of the species, variety or clone being considered. A sample of end-use requirements are those that

(1) best fill the end-use need (e.g., wood, alternate products, etc.),
(2) grow well on the sites in question or be suited for many sites,
(3) perform a range of ecological tasks,
(4) has useful secondary properties (e.g., flowers that attract honeybees or ants), and/or
(5) are complementarity with other species (i.e., as in a multi-species plantation or has evolutionary-accredited membership in native forests).

Formal selection is a matching process, on one side are the desired end-use requirements, on the other are those desirable properties (the DPCs as ordered below) that species possesses. In tropical and other highly biodiverse regions, there are tens of species from which to select a primary or a secondary component. DPCs methodologies reconciles the characteristics wanted and what is available in the candidate species.

Although the mathematical details are not part of this discussion (for a greater elaboration, see Wojtkowski, 1998), an ordering and matching of the DPCs eases the tangle in selecting from multitudinous candidate species. Even a few simple steps along this route, such as a determination of the DPCs for local tree species, can give better choices that may otherwise be possible.

Primary characteristics

Some generally stated, base requirements (the DPCs) for primary, wood-producing forestry species are:

(1) fast growth,
(2) tolerance to climatic variation,
(3) ease of propagation in systematized setting,
(4) resistance to diseases and plant-eating insects,
(5) straight true bole,
(6) good end-use wood properties,
(7) positive ecological attributes (e.g., fits well within a natural ecosystem),
(8) ease of handling (for planting stock), and
(9) self pruning.

The above are interpretive, the first five self-explanatory. The other requirements are perception dependent. Good wood properties can mean, e.g., (a) in pulpwood market, having long fiber and being easy to debark; (b) for woodworking, fine grained, decorative, nice color and figure, and easy to work; (c) for construction, physically strong, light, easy to work, and nails well.

Positive ecological attributes means that, in native ecosystems, the in-situ ecological attributes bring a species into ecological harmony with other, co-evolved plants, i.e., indigenous species presents fewer dangers to established ecosystems than do introduced exotic species. The hazards, and pros and cons, of exotic inclusion are discussed later in this chapter.

Ease of handling means that young trees proceed from the nursery to site with a high survival rate. This also means that it is easy to harvest (e.g., lacks spines on the trunk and/or branches or does not disproportionately dull saws).

Note that an ability to suppress weeds is not a general DPC. Weed suppression has utility but, as a DPC, this need is far from universal.

Secondary characteristics

The types of secondary species can be grouped by characteristics and/or use (i.e., as woody or non-woody; facilitative or productive). Whatever the grouping, there are desirable secondary characteristics that encourage use. Among these are:

(a) resource compatibility with the primary crop,
(b) an ability to grow on poor soils,
(c) tolerates climatic variation,
(d) ease of establishment,
(e) freedom from pests and diseases,
(f) lacks the capacity to become a weed,
(g) lack of root suckering (i.e., sprouting) properties,
(h) ability to trap nutrients,
(i) ease of control and eventual elimination,
(j) spinelessness (spines can also be a desirable property),
(k) a high rate of nitrogen fixation,
(l) dry season leaf retention (tropical plants only), and
(m) a preponderance of deep roots.

Beyond these, there are other, less needed, characteristics that fix a species in a secondary role. These can be

(n) ability to acquire nitrogen to the benefit of the primary species,
(o) rooting characteristics that do not interfere with the primary species, and
(p) the aforementioned weed suppression.

Complementarity

As purely mechanistic concept, plant-plant complementarity is presented in Chapter 3. There is more to this. There is the notion that plants have positive association with humans, animals, surrounding vegetation and natural ecosystems. Within larger landscapes, this may be what differentiates natural ecosystems from their agroecological counterparts. The concept of agrobiodiversity (discussed in this chapter) is built upon this.

Domestication

Most useful plants, including the common plantation trees, do not come directly from natural sources, but are filtered through some form of domestication. This process has different steps:

(1) an acquired knowledge on DPCs and growth parameters (site needs, best planting methods, intercropping potential, etc.);
(2) a search to find the best varieties;

(3) bringing to the fore the best genetic properties, i.e., to maximize growth and quality through cross-breeding and improved management; and

(4) expanding the climatic range in which the plant will grow.

Most of the agricultural crops have a long history of domestication. Maize and wheat have evolved from less productive, naturally-occurring plants to become of great economic significance. The number of commercial varieties of a given species is evidence of an intense domestication process. Some of the resulting variation can be less than obvious, rather than differences in color, shape, and/or size, differences can be manifested as a climatic accommodation, or drought or insect resistance.

For some trees, the end-paths have been agricultural. Walnuts and macadamia nuts reached market prominence by selecting (finding and cross-breeding) superior fruit, rather than wood-yielding, varieties. This often emphasizes short stature (for ease of fruit harvest), instead of long, clear stems.

Non-commercial, but useful trees, those encountered in intense farm landscapes, may have undergone a similar process (Lovett and Naq, 2000). In contrast, the common plantation trees of the world were selected because they have most of the recommended DPCs and a lot is known about their growth parameters and wood properties. A few, e.g., radiata and loblolly pine, have gone further down the domestication road through more rigorous selection and cross-breeding.

As seen, this is a specialized undertaking, not producing a universal end product. For trees, domestication aimed at service in monoculture can render trees unfit for insertion in polycultural plantations (Michon and DeForesta, 1997). This puts in doubt the ability of highly plantation-domesticated trees to thrive in bio-rich plantations.

Trees in intensely managed forests undergo a similar process (Finkeldey and Ziehe, 2004). Those of poor form are culled while holding on to the ability to thrive in a competitive environment. The varieties, rather than their plantation-domesticated cousins, may be best for a high intensity role in bio-rich plantations.

For all but a few tree species, domestication has not reached the first step, knowing enough to be able to rank the DPCs in relation to other species. Given the intricacies of the analysis, this step is not easily accomplished.

Native species

Throughout the world, most tree species have never been subject to domestication. There certainly exists trees that have superior DPCs but have not been identified as having such. This is especially true in tropical regions with a large number of species and limited knowledge on each.

This often biases selection. Teak is a southeast Asian tree that has gained widespread prominence. It has been extensively planted in Africa to the exclusion of local species. There is no reason why native African species, once better known, may have, or exceed, the DPCs of teak.

Exotic trees

What complicates domestication are exotic trees. As stated, these find application because local species lack an established knowledge base, because they grow better (growth rate and/or form), or they tolerate a climate or site better than local species.

Exotics can be relatively free from local pests (although exceptions exist) and this may encourage use. Rubber trees suffer major infirmities in native Brazil, but are less prone to disease as exotics in Africa and Southeast Asia. Other gains come in being able to expand timber production to sites unsuitable for native trees (as with drought resistant eucalyptus) or, through high yields, taking the productive pressure off native forests (Peterken, 2001).

The disadvantages of exotics, as modified from Peterken (2001) and Williams (2002), are:

(1) a reduction in overall biodiversity through
   (a) the loss of ecosystems through ecosystem change or site degradation (e.g., soil drying),
   (b) non-compatibility with and the possible loss of rare, local plant species,
   (c) a reduction or elimination of local fauna (e.g., through the loss of feed and nesting sites);
(2) allowing more invertebrates to become pests;
(3) contradictory hydrology (reduced water runoff or groundwater);
(4) introduced fire danger;
(5) aesthetic diminishment.

Numerous examples illustrate what can happen when exotics clash with local conditions. Ohte et al. (2003) observed that, in semi-arid China, two local species, *Sabina vulgaris* and *Artemisia ordosica,* survived by conserving water while an introduced species, *Salix matsudana,* freely soaked up groundwater. The overuse of water is not the best DPC where shortages loom. In a less water-constrained site, planting non-native pines (e.g., *Pinus caribrea* and *P. patula*) excluded native Africa trees (Struhsaker et al., 1989).

Eucalyptus, outside native Australia, is the most commonly cited tree species with site-contrary DPCs. Among the drawbacks, this species has been faulted (a) for a large water intake, changing the hydrology of the land; (b) for non-compatibility with local flora, negatively altering local ecosystem dynamics; and (c) for increased fire danger (Williams, 2002).

Purposeful introductions are only one part of the problem. Introduced trees can become weeds, thriving and excluding native species. Notable cases have occurred in Australia and New Zealand where large tracts are unwittingly taken over by introduced plants. This has been done with lodgepole pine in New Zealand (Ledgard, 2001).

Many invasions are less noticed, especially in complex ecosystems. Turner (1994) mentions that, on the west coast of North America, introduced species, those that sprouted quickly after a fire, reduced the number of beneficial native plants. These can influence successional dynamics.

Complicating matters are those instances where a tree species does not grow well where native. The explanation is that subtle climate or some other change has come about faster than these species can evolve. As a consequence, some may grow better in other regions.

This is not as farfetched as one might suppose. A commonly cited example is radiata pine. This has become a leading plantation species in the southern hemisphere while only a passing curiosity in its native territory of southern California and northwest Mexico. A lesser known case, klinkii pine (*Araucaria hunsteinii*), native to the highlands of New Guinea, no longer thrives where indigenous.

ECOSYSTEM PERSPECTIVE

Trees are grown in stands which, by definition, contain a single, more or less uniform agroecosystem with a set silvicultural treatment.

A stand, especially inside a natural forest, need not be geometrically derived (i.e., be square or rectangular), but instead conforms to soil and/or topographical norms. Plantation stands can be square or rectangular (geometric) or have an irregular outline set by land contours, water flows, soil types, or some other site characteristic.

A part of this discussion involves the degree of intended agroecological interaction within a stand and with neighboring areas (Wojtkowski, 2004). A single stand might be a series plot with trees in varying successional phases. These are considered a single stand because, as management units, these are cross-linked by various ecological influences.

What occurs outside a stand can be a delimiting influence. This can occur when nearby seed trees re-establish a stand after a clearcut. In this case, the bounds of the stand are set by, and include, the adjoining seed trees.

These influences radiate out. A collection or group of stands, often placed for specific ecological purpose(s), constitutes a landscape. The landscape can be series of plantations, a multi-stand forest or some farm-forestry construct. Stands or landscapes have no intrinsic size limits. These are set by outside factors.

Component Rankings

In species-based systems, plant species are ranked as primary or secondary. This is more of an economic measure. Nature has other ideas and, in complex ecosystems, a species can be member of dominant, co-dominant (or sub-dominant), intermediate or suppressed class (Moon and Brown, 1914).

The dominant trees, of one or more individual species, are those that, because of size, have greater access to and take a larger percentage of the essential resources. These can be defined as receiving vertical and horizontal light.

In untouched tropical rainforests, some species that are dominant in the upper canopy and a few that are emergent. The latter are those scattered trees that clearly rise above the upper canopy. Emergent trees might also be considered as super-dominant.

Co-dominants are slightly smaller in height and capture a greater percentage of vertical light. These still have enough presence to maintain their share of essential resources and compete actively against dominant species. Both the dominants and co-dominants occupy, either through size and/or aggressiveness, the most desirable niches.

Within gaps in the uppermost canopy are intermediate trees. These fend by catching vertical light and capturing resources missed or not needed by the dominant and co-dominant species.

The suppressed class, those under the others, must get along by exploiting narrow niche openings or by utilizing essential resources missed by the taller species. These may be suppressed, but many have future pre-eminence and commercial value.

Trace species are less noticed in forests. These are found in very low, per area densities but, because there may be a lot of these, they can, in total, exert considerable ecological influence.

Mostly herbaceous trees also enter in this category. In tropical forests, there can be a considerable distance between like species (Pitman et al., 1999), so dominant, co-dominant and intermediate trees may also be trace species.

Agroecosystem Types

Silvicultural ecosystems have distinct ecological and economic dynamics and with stable (and desirable) properties. Agroecosystems can be classified through varying criteria (Wojtkowski, 2002);

(1) types of outputs (tree species, crops, animals, etc.);
(2) economic orientation (revenue or cost);
(3) temporal periods;
   (a) of the component species,
   (b) overall system rotation,
(4) the agrobionomic principles employed (facilitation, suppression, etc.);
(5) landuser objectives; both in
   (a) productivity of the primary and secondary trees,
   (b) the land-use problem(s) resolved,
(6) benefits received (both tangible and intangible).

One of most basic agroecological divisions whether systems are formulated to produce useable output or whether they serve a non-productive function within a landscape.

Where wood output is expected, this is a principal-mode system. Where an ecosystem serves a specific ecological role and output, if any, is secondary, this is an auxiliary system. In silviculture, these two subdivisions are not as prominent as in mixed agriculture-forestry situations, but the concept does have application.

Principal-mode

Principal-mode systems are productive units and, in forestry, these are the wood-producing ecosystems. Blocks of plantation trees clearly fall within this category, as do the mixed tree-crop systems of agroforestry or blocks of natural forest in an otherwise agricultural landscape.

Auxiliary

Although the applications for non-producing ecosystems are limited in silvicultural situations, these still exist. Within this grouping fall riparian systems, shelterbelts, windbreaks and other applications where wood is secondary to the ecological gains in having a specific-purpose ecosystem, one with predetermined ecological properties, at a set location.

Objectives

The objectives of ecosystems and landscapes are many and varied. Specific purpose ecosystems can counter a range of scourges (drought, flooding, insect and disease infestation, etc.). Planned ecosystems can also bring productive systems more in harmony with nature. This is done by offering habitat for birds and canopy and ground-dwelling animals and partial refuge for certain classes of vegetative biodiversity.

Objectives extend to forest ecosystems. Forests are a source of wood, provide habitat for natural fauna, serve as watersheds and have recreational possibilities. A forest can be multi-purpose, offering all these or a forest can be designed to encourage one purpose, while maintaining others. For example, while being managed for silviculture, recreation, through trails and vistas, heighten the natural experience, adding another purpose (recreation) without lessening others (wood output, fauna habitat, etc.). Species of animals can be encouraged or discouraged by modification in habitat. A forest enriched with key tree or plant species will favor those critters that are linked with the plant species (one or more) being added.

Desirable (Agro)ecosystem Properties

The wood produced may influence, but does determine the type of ecosystem derived. In addition to being a productive unit, the landscape also has ecological, economic and environmental purpose and this can be enhanced through type (properties) and placement of individual ecosystems. These are selected and placed on the basis of desirable (agro)ecosystem properties (DAPs). A sample of DAPs includes being:

(a) conducive to high tree growth rates;
(b) resistant to erosion;
(c) offering drought and flood immunity;
(d) unsuitable for herbivore, root and/or wood attacking insects;
(e) uninviting for the spread of plant diseases;
(f) wind resistant;
(g) while not encouraging damaging wildfires.

This concept is akin to that of DPCs. Instead of individual species, the characteristics of entire ecosystems can be listed and ranked as to site and labor needs, sustainability factors, environmental concerns, and nuances of practical application.

As with DPCs, there are two sides to the equation (1) what is wanted from an ecosystem (the objectives) and (2) what an ecosystem can deliver (the DAPs). The essence of the DAP approach is in matching as close as possible these two lists. Although the concept is simple, the devil is often in the detail.

Design

Confirmed properties of agroecosystems can be ranked and an appropriate selection made on this basis. This idea can be channeled further and agroecosystems can be designed for specific ecological purpose or with a specific DAP rankings.

Known silvicultural practices are a starting point as these have productive purpose, overcome specific land-use problems, and address discrete socioeconomic need(s). In being better understood, the known effectiveness that a design (spatial, temporal, etc.) has on DAPs helps guide the process. The more intrepid and knowledgeable may forego these as standards and formulate a system from scratch. The design variables are pivotal in this process.

Variables

DAPs can be a function of the management tools used. Some have been described in previous chapters (e.g., spatial and/or temporal pattern), others follow. There is some latitude to substitute an active intervention (i.e., manual labor or chemical inputs) to one controlled and/or undertaken through natural dynamics. The design variables used in achieving set goals are:

(1) component species
  (a) primary (one+)
  (b) secondary (one+)
(2) spatial ordering
  (a) spatial pattern (horizontal)
  (b) arrangements
  (c) interplant distances
  (d) canopy patterns (vertical)
(3) temporal pattern
(4) management options/inputs
  (a) labor inputs
    (i) weeding
    (ii) thinning
    (iii) tree establishment (planting)
    (iv) tree pruning
  (b) supplemental inputs
    (i) fertilizers
    (ii) insecticides
    (iii) herbicides
    (iv) irrigation
    (v) root barriers
  (c) stem/branch grafting
  (d) planting or tillage method
  (e) fire
  (f) grazing

Principles

In designing an intense forestry ecosystem, the productive purpose need not be subverted. Once the primary species is set, DAPs arise from secondary species (one or more), their placement within the ecosystem, the placement of ecosystem within the larger landscape and the management options.

One unproven, but plausible hypothesis, is the dichotomy between species composition and spatial layout. Some DAPs, those not achievable by composition, may be introduced through spatial pattern. In reverse, those not obtainable through pattern may be initiated by species. If both fail in deriving the desired set of DAPs, then the management options and/or the landscape associations may prove helpful. The temporal patterns, as discussed in Chapter 4, also enact a wide variety of options and outcomes.

Variable use

The basics of any design are set by the component species and the spatial and/or temporal pattern. Beyond this, the management options determine the economic and ecological feasibility and viability of any design. There are those management options that are key, without which the system cannot find direction and use. The others are the slack variables which, although implemented, do not contribute to the success of any design.

With grazing under pines, pruning can be the determinate variable as this opens the canopy, encouraging better forage growth, while ensuring stem form. The planting method can also be key where, for example, large stem planting can reduce the rotation and improve log form, contributing to the success of the design.

In contrast, pruning is not a concern in some plantations. Excessive branching may be controlled through manual labor but, as a slack variable in a well thought out design, this is redundant as self pruning is an ecologically installed mechanism (as discussed in Chapter 6).

Optimization

Inherent in design is the concept of single or multiple objective optimization. If an ecosystem is to be formulated for a specific purpose, the best possible design is advisable. This is easier where wood production is the single objective. The silvicultural treatments presented in this text have wood output as the utmost objective.

This is not always so. Wood may be important, but not uppermost. Without a single objective, design becomes a multiple objective problem. This is common and usually handled informally by those overseeing the situation.

There are formal, if undeveloped mathematical techniques, for solving multi-objective land-use problems. These are often curbed by a lack of data or difficulty in translating known variables into dependable numerical equations. Despite this, there is value in looking at how these can be solved

Multi-objective optimization can (1) rank objectives according to their importance or (2) be single objective with tie breaking. The first, the ranking of objectives, tries to formulate a compromise where all concerns are partially addressed. The compromise may be tilted in favor of those of greater importance. The second, tie breaking, looks at all those land-use alternatives that maximize the most important objective. A second objective comes into play where there are a number of designs that satisfy the first concern. If so, the second is the tie breaker. If multiple design choices still exist, a third objective comes into play.

Agrobiodiversity

Extracting the most from any one ecosystem may depend upon biodiversity (Naeem et al., 1994; Kareiva, 1994). Even forestry monocultures may gain much from the unnoticed micro-organisms and casual, non-intended, vegetative inclusions.

With designed and managed ecosystems, all intended species should have a purpose. This is agrobiodiversity, where species-richness is judged by the role or purpose that each species has in a planned ecosystem and/or how a plant is utilized by local peoples and communities (Jain, 2000). With an agrobiodiversity approach, all plants should have a purpose and there should be many of these.

A pure agrobiodiversity approach has dangers in planned and managed plant communities. Wild plants, those that have no apparent productive or facilitative function, can be viewed as extraneous and not worthy of attention and inclusion (Simberoff, 1999). An ecosystem design, as intended, may limit the amount of natural (casual) diversity tolerated and present, however, optimum results can be obtained with a large percentage of agrobiodiversity, however, with considerable casual biodiversity.

Despite the drawbacks, agrobiodiversity or directed biodiversity is part of a larger picture. This relates to landscape agroecology, community forestry and other aspects of the people-nature interface.

Mimicry

Some consider the best land-use practices are those that mimic what occurs naturally on a site (Lefroy et al., 1999). A number of agroecosystems have this property, rice paddies mimic, in some aspects, swamp ecosystems; parklands (pastures with scattered trees) duplicate, in part, the ecology of natural grasslands; and agroforests can be the equivalent in biodiversity and landscape role to natural forests.

The goal of mimicry is to closely imitate what nature intended without seriously compromising the productive role of the ecosystem. In silviculture, the option exists to plant forests where needed and exploit these so that natural flora and fauna can survive or even flourish. Some guidelines for implementing this view are to

(a) curtail the inclusion of exotic species;

(b) encourage agrobiodiversity, especially in lesser native plants;

(c) accommodate or tolerate incidental biodiversity (plants and animals that arrive, unannounced); and are most influential;

(d) match the tree species (more than one) and the ecosystem (through, for example, especially derived DAPs) with what nature has intended for a site (e.g., a closed canopy ecosystem, groups of trees interspersed with open ground, etc.).

CHAPTER 6

Niche Transitions and Ecological Services

The previous chapters discuss basic ecosystem formulation principles. A not-to-be-overlooked aspect of agroecology is in instituting niche transformations and natural ecosystem dynamics to overcome adversities (droughts, insect infestations, disease spread, etc.). Similarly, ecosystem dynamics also provide management, reducing costs by replacing labor and chemical inputs.

Revenue can be lost and costs increased fighting what nature has bestowed. Best results are obtained managing a stand and site as nature intended. In Chapter 4, vegetative-site improvements, as part of a successional strategy, are briefly presented. These are clear situations where a site-improving, facilitative species, planted in anticipation of high value tree, can make a site more accommodating for upcoming species (one or more).

Along with site modification, planting, thinning and weeding are all management activities that can be done expensively by hand or, if an ecosystem is well formulated, are a free service provided by nature. Overlooked to the detriment of the landuser, these go to the heart of nature-aided agroecology. Some of nature-induced management options and ecological tools are overviewed here; refined in subsequent chapters.

GENERAL GUIDELINES

All ushered in plants, whether exotic or local, are subject to general guidelines. From Schenck (1904) and Schlich (1910), the attendant do-no-harm admonitions are:

(1) having a locale favorable for the species planted; and

(2) where the species should not deteriorate the site.

Clearly, planting trees that are site-suitable is overriding. Productive, as well as environmental gains, come in utilizing the most site-adopted tree species (generally a mix of species is better). Harm occurs when the tree chosen precipitates less favorable soil structure and composition, impoverishes local hydrology, enfeebles local flora and fauna, and/or transforms the area micro-climate.

As shown with examples in Chapter 5 and through the discussion of species co-evolution in local ecosystems (Chapter 2), the blame for ecological harm can lie with miscast non-native species. Some exotics fit, others do not. It may be for this reason that there is disagreement on the environmental risk, in magnitude and/or importance, from introduced trees, e.g., Zobel et al. (1987).

Even if the site-contrary argument is accepted, this does not mean that wood producing, regionally-transplanted tree species should be totally avoided. Given the trees available locally, people may have to look outside, at non-native species, for needed DPCs.

In the case of eucalyptus, the straight clear stems, rapid growth and drought tolerance make this an attractive silvicultural addition. Few trees have this combination of DPCs. This species is faulted, among other reasons, for aggressively seeking water, denying other species, negatively impacting the surrounding area.

If introduced, it should be on terrain where the potential environmental damage is limited, i.e., (a) the hydrology is less on the margin; (b) little else grows because of moisture limits and there is little to disturb; (c) the rainfall patterns favor thirsty trees with little negative consequence; and/or (d) due to placement, an unsatisfactory effect is not widely transmitted, such as planting on dry hilltops.

Exotic species are not the only source of concern. The over planting of a local species, especially to the exclusion of others, can upset regional ecology and violate the do-no-harm admonition.

SITE ACCOMMODATIONS

Silvicultural considerations are history dependent and can involve reestablishing what once was. In regions that have experienced the complete removal or destruction of forests, direct re-establishment may not be possible. Drying on wind swept plains; nutrient loss;

animals, wild or domestic, eating new vegetation; or a break in the rainfall pattern can create a terrain inhospitable for trees and forests.

Examples go back to antiquity. The Mediterranean region was once far more forested than today (Hughes, 1994). Human activity has greatly reduced this. Uncontrolled grazing and unwise wood extraction are often blamed. A temperate study is found in the Patagonia of Chile where grassland replaced forests burned in the early 1900s. After many decades, a fair percentage of the area is still tree-excluding grassland.

Tropical rainforests exist, not through a rich soil base, but through the continued internal cycling of essential resources. Breaking this cycle and/or the concurrent rainfall patterns (Yoon, 2001) can result in slow or non-reestablishment.

Preconditions must exist before trees and forests occurs. This can require vegetative and/or physical site modifications to bring or keep soils in a tree-receptive state and/or to counter inhospitable events; those that obstruct or hinder growth.

Water may be foremost on a comprehensive list, but excessive wind, poor soil and grazing or wildlife are included. Human activities, e.g., unconstrained firewood collection and the pasturing of domestic animals, also hamper tree growth. These can be overcome.

Physical Changes

In agronomy, the land-use intensity makes physical modification economically tenable. Modifications include draining swamps, building terraces, and constructing paddies. In forestry, the cost seldom justifies intense land modifications although, in scattered situations, less dramatic measures may be taken. Moisture around individual trees can be increased through (a) micro-catchments, (b) rock piles, or (c) infiltration ditches.

Micro-catchments are small diversion ditches that direct rainwater to the root structure of a planted tree. When rainfall is marginal to establishment, these structures allow trees to gain a foothold (Carucci, 2000). When micro-catchments were employed in arid India, Gupta et al. (2000) found tree survival increased from 50% to 90%. The cost of planting also increased 20-30%.

Rock piles are less known and involve planting young trees in clusters of small rocks. The rocks provide (a) a more porous soil surface, (b) retard soil drying around the plants, (c) moderate nearby air and ground temperatures, (d) discourage some small animals, and (e) attract windblown nutrients (Fish, 2000). Steenbergh and Lowe (1969) describe the higher survival rates in rocky, as opposed to seemly more inviting alluvial sites.

Other measures are larger in scale. Infiltration ditches (and/or bunds) contour hillsides, serving the same function as the micro-catchment. Trees are planted at wide intervals along the bottom or side of the ditch. Filling the ditches with organic materials (leaves, decayed branches and stems and other water holding materials), retains more moisture near newly planted trees.

Vegetative Modification

Where conditions warrant, it is contingent upon establishment practice to set the regional stage for the return of trees and the forest. This can be a one or two-step process.

The first step can be in finding a commercial tree-species that is capable of overcoming adversity on a now inhospitable site. Many practitioners are content to remain in this stage, selecting a species that does not encourage further ecosystem development, e.g., species of eucalyptus are commonly employed where moisture is the limiting resource.

Others may opt for more, employing woody and/or non-woody species to set the ecological stage for the return of native trees or forests. The pathfinding species can be a commercially viable tree or a purely facilitative plant. The key DPCs are in (a) resource compatibility with the future ecosystem, (b) being able to overcome an inhospitable site, and (c) resistance to growth impediments, e.g., grazing.

A common form of vegetative pathfinding has nitrogen-fixing plants prepare the soil for an upcoming nitrogen-demanding tree. While not being the only possibility, there is the already alluded to, phosphorus-accumulating while tree species *Tithonia diversifolia* (Buresh, 1999).

These are not the only implementations of a very useful agroecological tool. In western North America, a mix of rye or winter wheat and natural grasses are seeded after a severe fire; protecting soil resources from heavy rainfalls. The rye and wheat establish rapidly, the grasses quickly follow, trees come in a second round once the site is stable.

Vegetation is often employed to restore badly eroded sites. In this form of land repair, specialized plants both steady the soil and increase the nutrient and water-holding capacity. This may set the stage for future usage or be simultaneous with some forestry application. Photo 6.1 is of a badly eroded site being stabilized by shrubs.

Part of the restoration lies in protection, mainly micro-climatic, furnished by pathfinding plants. These retard evaporation, allowing growth where not possible without this facilitation mechanism. Another facilitative gains may be afforded directly through hydraulic lift or indirectly through the introduction of a micro-ecosystem.

Mineral concentrations (as with toxic soils) that inhabit plant growth can also be countered. Bio-remediation is accomplished through plants that (a) absorb an unwanted toxic mineral or (b) buffer the concentration. Both leave the soil more inviting for other, less tolerant, species.

Photo 6.1 Shrub vegetation being employed to stabilize an eroded site making it suitable for a future, higher-valued use.

Less well documented, but still accepted, is the notion that rainfall is a function of a vegetative cover and, with more on-site living biomass, greater rainfall can be expected. Not applicable to a few plots, landscape wide, there is evidence to support this premise (Sandström, 1998; Ataroff and Rada, 2000).

Progressive Plantings

Isolated trees are found growing in trying locations. When these sites are very inhospitable, tree growth is a cause for wonderment. This may be in a desert where a single sequestered tree begs the question as to how the tree was able to establish and grow in such a forbidding area.

The answer lies in a combination of fortuitous events that have a seed in place at the beginning of some wetter years. Continued survival is contingent on the tree striking below ground moisture within the wetter-than-average year. It helps if the species is highly drought-resistant.

In semi-arid zones, it is not unusual to replant the same plot across many seasons, all in the hope of a wet year and an acceptable rate of tree survival. Nature has the luxury of time, silviculturists may not. If continued replanting in the hope of a few good rainfall years is not a workable solution, mixed temporal and spatial progressive plantings may be. Photo 6.2 shows a drought-resistant tree species on a high-risk site.

Spatially progressive plantings are initiated along the bottom of moisture-rich wadies. Once the lower elevation trees establish, the planting proceeds uphill in yearly or seasonal increments. The hope is that, through micro-climatic protection and/or hydraulic lift, the newer plantings in each progressive step will have an improved chance of survival. The temporal aspect comes as not all the trees are planted at once, instead the process spans many seasons.

Micro-catchments, bunds and similar structures can anchor a progressive series. Generally, the larger scale, water-slowing structures are constructed first at higher elevations, progressing to lower reaches. The idea being that, if first placed at lower elevations, infiltration ditches or bunds will washout during high rainfall periods. The smallest structures (the micro-catchments) have more locational flexibility.

Photo 6.2 A drought-resistant acacia species with the natural ability to overcoming the limitations of a high-risk site.

In a flatland version of the progressive planting, shelterbelts and windbreaks, established at wide intervals can change a dry, windswept plain into a site more hospitable for tree stands. The idea being that, once the initial plants are established, the space between is exploited.

Among the options are to (1) establish windbreaks where trees easily grow, e.g., along streams or wadies, and fill the voids once conditions permit or (2) employ a drought-tolerant exotic tree as a windbreak, e.g., a eucalyptus or an acacia species, to set the stage for less tolerant-native species. With the proper design, a windbreak or shelterbelt also contributes, though wood output, to the overall economic yield. If the improvements are sweeping enough, a forest may eventually form.

Countering Grazing

Another common problem are grazed areas; recently planted trees are in danger of being eaten by domestic or wild animals. Small trees are subject to other animal-related risks, such as being stepped on by large mammals.

Grazing is especially problematic in arid zones where the populations of wild or domestic animals may be large and, through a long pronounced dry season, feed is in short supply. This is not unexpected. In some natural ecosystems, populations of wild animals eat up to 80% of the biomass (Vera, 2000).

The benefits of abstinence are clear; trees thrive if an area is kept free of animals until tall and the leaves are out of reach. Where not possible, there are defensive measures, i.e., through species selection, planting method and through animal management.

Grazing animals find some species less palatable than others. The leaves of pines and eucalyptus, possessing less tasty, internal chemical compounds, are less desirable as food. A few plants, e.g., many African acacia species, are well protected by thorns. Whatever the case, a species can be planted that animals do not readily eat.

If the planting method (next section) fails to alleviate the danger, animal barriers can be employed. When first planted, seedlings can be surrounded by leafless, thorny bushes. For taller trees with out-of-reach canopies, the bark can be guarded by (a) very small diameter branches tied around each stem, (b) by plastic tube guards, or (c) by painting the trunks with a repellent. A number of repellent chemicals have been utilized; the cheapest and most available is animal manure (Gill and Eason, 1994).

PLANTING METHODS

How trees are planted has a direct bearing on the type of silviculture practised (or even if trees can exist or be considered). As mentioned, on some sites, accommodation must be made and non-economic adversities overcome before significant growth is achievable.

The planting options are yet another tool in this. Where trees ecologically befit a site, planting methods may be more in response to economic need; planting costs, available machinery, rotation period, etc.

Nature employs seeds and sprouts for reproduction, but silviculturists have a wider array of options, each having advantages, disadvantages and specific applications. The planting methods are: (1) seeds, (2) seedlings, (3) striplings, (4) stumps, (5) sprouts, and (6) stems.

Seeds

For a silvicultural field planting, seeds are often utilized. Difficulty in the planting process, especially with very small seeds, low survival, genetic uncertainty, and longer juvenile period are the primary disadvantages.

This method is best where large, remote areas must be quickly planted and quality is less important, e.g., after an expansive fire. Short-lived, seed-instituted herbaceous species are commonly established in post-trauma situations. The trees of commercial interest come at a later planting and are usually not seed established. Natural regeneration in natural forests can be mostly a seed-based affair.

Seedlings

In many temperate regions, seedlings are the most common planting method. Usually the trees are at a height of 10 to 20 cm before being taken from a nursery and placed. The advantages, as compared with seeds, are that (a) individual quality characteristics can be assessed before replanting, (b) survival rates are improved, (c) the rotational period is slightly shortened, and (d) the seedlings are easy to handle in the replanting process. Among the disadvantages are a requirement to protect young plants from being grazed or stepped upon by larger animals.

Seedlings have risen to prominence simply because a lot of plantation planting utilize pines. This is somewhat unfortunate as, in conforming to the planting requirements if the pine, the other planting methods have faded from view.

Striplings

The problem of grazing spurred the utilization of striplings. These are long-stem seedlings, tall enough to be out of the reach of shorter animals. Generally, the trees are 1.5 to 2 m high when replanted. The lower branches are removed leaving only a small terminal canopy. This eases handling and improves survival (by reducing transpiration while the disturbed roots develop).

The disadvantages, as compared with seedlings, are a longer in-nursery period, increased difficulty in handling, and the added labor involved. Besides protection against marauding animals, this method provides better planting stock as only trees of superior stem form are transplanted.

Stumps

Stump planting is an alternative to seedlings. For this, plants of about 4 cm diameter (measured at groundlevel) are taken from the nursery, cut about 2-4 cm above the ground line and from 10-25 cm below. The resulting stump is easier to handle, transport, and to plant, especially in remote regions. Planting is accomplished by poking a hole in the ground with a long heavy (often metal) pole and the stump placed in the hole.

In addition to survival gains, stumps provide vetted stock, more so than with seedlings. The disadvantages are climatic (i.e., restricted to wetter regions), with species (i.e., confined to those that sprout readily), and where weeds establish and overwhelm quickly. This method does not work with pines, spruce and other coniferous species.

Sprouts

Sprouts are seldom employed in new plantings, but can be a mainstay as a re-establishment method. In the lesser application, uncut branches from living trees are laid upon the ground and allowed to root. This also has relevance, mainly to fill gaps between existing trees and can be a regeneration method in multi-aged stands.

More common silvicultural usage is, after trees are harvested, with stumps that re-sprout (i.e., coppice). This can shape a new plantation or the regeneration in a natural forest. In either case, the primary advantage is cost, although some labor is needed to prune the multiplicity of stump sprouts. The main disadvantage is that planting stock is not genetically upgraded upon re-establishment.

Stems

Stem planting comes in two forms. The easiest is branch stock which is cut from nearby trees, carried, and planted as a leafless stems. Planted branches are generally about 1 m in length. The elimination of the nursery is the chief advantage, but survival and good tree form are not guaranteed. This method may be employed where the work must be done quickly and conventional planting stock is not available. Due to survival concerns, this may be limited to wetter regions.

The second method engages the main stems of trees. The long-stem planting method is underused and finds application with very high intensity, short rotation situations. For this, nursery stock reaches 5 to 10 m in height. The stem is cut at ground level and the leaves and branches removed. The entire stem is planted to a depth of 1 to 2 m.

The advantages are: (a) subsequent log quality is assured; (b) if well planned, there is a high survival rate; (c) there are fewer problems with grazing; and (c) the rotational period can be reduced by 10 or more years. The disadvantage is the long in-nursery period and the need for favorable terrain and mechanized planting. Photo 6.3 has a stand, newly planted using long stems.

It should be noted that the higher planting costs may be low in relation to the economic benefits received. Besides greatly shortened rotations, immediate grazing is possible and weeds are far less of a concern. Neither of the stem planting options works with pines and other coniferous species.

Photo 6.3 A long-stem tree planting. In this example, the trees have been in place for only a few weeks. The height (about 10 m.) serves as protection from grazers and to reduce the overall rotation.

WEEDING OPTIONS

Once all is in place, other factors intercede to insure or impede silvicultural success. The control of weeds, especially in the early stages of a plantation, can be of paramount importance. Weeds can include natural regeneration from a previous forestry land use.

In some situations, the critical nature of weed control cannot be understated. In the humid tropics, with high growth rates and an abundance of weed species, a poor planting sequence means that a site must be abandoned until the weeds subside and conditions exist to restart the sequence. Resumption can be some years in the future.

Although a greater threat during planting, weeds do negatively affect standing trees. Total elimination may not be the best option and not all weeding schemes rely upon total eradication. As a topic more applicable to the monoculture, the hand removal options are detailed in Chapter 8. Other control methods, those that put the mechanisms of exclusion (Chapter 2) into practice, are discussed here.

Natural Control

It is certainly far better if nature controls weeds without the need for hand weeding. The mechanisms of exclusion are put into practice through ecosystem design. The influences sought are to

(a) ensure that most essential resources, especially those considered limiting, are appropriated by the primary species (one or more);
(b) accommodate a primary species that actively discourage weeds; and/or
(c) add secondary, purely facilitative species (one or more) that can sop up excess resources.

The first of these, a resource demanding ecosystem, comes into play with greater agrobiodiversity (i.e., biodiversity in primary and economically-active species). The ability to occupy multiple niches is a corollary of the multi-species plantation and, as such, discussion continues under bicultures and three-plus polycultures (see chapters 9 and 10).

To discourage unwanted plants, a single tree species can be planted which does just that. Chapter 2 presents the mechanisms of species-based exclusion. This can be a strategy of the monoculture or a multi-species planting. Highly weed-competitive species include many species of pine, that once having gained site-supremacy, require little maintenance. Pines species must still past through the weed-competitive shrub phase to reach the maintenance-reduced, early successional phase.

Having a facilitative plant sop up excess resources is viable strategy and, in silviculture, cover crops have a long history of use. Typically planted are herbaceous species that produce a thick, ground-hugging mat; blanketing the ground and crowding out weeds. This works best in the scrub phase, but may be the mainstay of weed control strategy for weed-susceptible trees in subsequent phases. The DPCs of ground-hugging cover crops are presented in Chapter 9.

Other plants and other mechanisms are possible. Large, slow decomposing leaves provide a measure of weed control (Budelman, 1988; Eason, 1991). Therefore, short-statured, non-competitive trees and shrubs, those that shed copious amounts of broad leaves, are potential cover crops. From a wider perspective, any species, trees included, that can occupy unused niches, coexist with the primary species, and block weeds, can serve as a cover crop.

Fire

Fire is beneficial tool, both in stand management and in overseeing larger ecosystems. In the natural forests of western North America, periodic fires remove the undergrowth, encouraging the retention of existing climax species, and prevents highly destructive crown fires.

Fire can be employed to promote desirable vegetation. In a case study from western Canada, Turner (1994) found fire increased berry crops such as red elderberry, white flower current, thimbleberry, and swamp gooseberry. This had a positive affect on tree growth, overall biodiversity, and on the economic usefulness of the system. Two applications are discussed here, fire for (1) stand management and (2) land clearing.

Land clearing

With land clearing, fire serves both to remove old vegetation and kill aspiring new weeds. The idea being to skip the scrub phase of a natural succession, proceeding to a plantation or other silvicultural activity.

A hot fire is usually advocated. When dry vegetation is burned after a long dry spell, the high temperatures will kill weed seeds in the humus layer. The disadvantage is that more nutrients are lost through burning and unprotected soil can be lost through subsequent erosion.

A hot, land-clearing fire is usually followed by tree planting or, in a two-step process, the planting of a cover crop followed by trees, as with the rye, wheat, natural grass example. If natural regeneration is the goal, the burned areas are kept small, reducing the magnitude of the danger and encouraging subsequent reseeding from nearby untouched areas.

Stand management

The reasons for fire as a stand management tool are (modified from Shaver, 2002):

(a) to prevent fuel buildup (fallen stems, branches and bark) and large, tree damaging burns at a future date;
(b) to reduce the number of unwanted competitive or less useful plants (including pine regeneration in pine plantations);
(c) to help control insect and disease;
(d) quickly recycle the nutrients in post-harvest tree residues;
(e) increase natural biodiversity in fire-evolved ecosystems.

In short, continued use can bring on relatively species-harmonious, fire-resistant ecosystems. This can occur in plantations as well as natural forest settings.

Normally, one seeks a well-timed, cold fire, one that passes quickly, burns little, and kills surface weeds. The coldest fires are usually carried out after rains before the vegetation is fully dried. This is easier in a plantation with a single, fire-resistant species. Lacking fire resistance, a small area cleared around each tree permits a cold burn with little harm.

Timing can be important, not only to accommodate weather and control the magnitude of a burn, but to target certain classes of vegetation. The idea is to burn unwanted woody perennial and herbaceous weeds when these are less likely to survive. With woody perennials, the likelihood of survival is greatest when stored resources are high (as in the early spring or after a dry season) and the plant is getting ready for a growth spurt. Survival is lessened immediately after a growth period when internal resources are low (Malarson and Trabaud, 1988).

Other applications of fire require considerable finesse. Fire can be implemented to thin unwanted natural regeneration, leaving the taller survivors with a more economically advantageous spacing. This is a task not undertaken lightly, success requires considerable experience. Done right, the resulting stand is disarrayed and clumped, but with a spacing closer to that needed for optimal growth.

In addition to weed removal, fire can change the composition of the understory and future ecosystem. In fire-affected, event-driven ecosystems, the type of blaze can alter site niches in favor of one species or another.

In an example from Arizona, Barton (1999) found that moderate intensity burns favored pines, while a less intense fires benefited oaks. Brose et al. (1999) employed fire as a tool to alter the successional dynamics from yellow poplar to oak. Altering successional dynamics through fire requires a knowledge of stand dynamics, a sense of direction as to what the result will be, and an understanding of the environmental stresses that might be imposed (Huddle and Pallardy, 1996).

Grazing

Under establishment, the protection of young trees is briefly discussed. Although grazing animals are a danger during the establishment phase, they can be a useful tool in plantation or woodland management once the tree canopies are beyond reach (giraffes being the notable exception). As shown in Photo 6.4, eliminating or reducing undergrowth (weeds and competing young trees), the taller trees experience better spacing and growth (Kienast et al. 1999). Grazing also helps retain or bring in essential plant nutrients (DeMazancourt et al.,1998).

With foresight, this tool can be refined. The type of animal, either domestic (cattle, goats, horses, etc.) or wild (deer, lamas, bison, etc.), is one variable. The others are the stocking (per area) population, the age of a stand, and amount of time these animals are permitted to work.

As tools, foraging animals can be divided into three categories (Hoffmann, 1973):

(1) browers which eat leaf buds, leaves, twigs, and bark (examples include the moose and giraffe);

(2) grazers which eat grass (examples include sheep, cattle, and plains bison); and

(3) intermediate feeders which eat both grasses and woody plants (examples include goats, some deer species, impala, woods bison and gazelle).

Photo 6.4 A grazed forest where the understory has been eaten.

Generally, the animal is paired with the types of weeds needing removal. Within the above groupings, sheep and cattle prefer tender, low-growing vegetation; goats eat coarser vegetation starting with taller plants; horses can reach and consume tree branches.

In theory, if a tall coarse weed needs to be eaten, animals that like tall coarse weeds (e.g., goats) are employed. If precise animal species selection is not possible, then those animal species that are available are directed to target plants through the stocking rate and/or timing.

Grazers eat their favorite foods first, later going on to less liked species. If the targeted species is less well liked, it will take more time for the animals to eat it. Some situations are best handled with very high populations so that the animals quickly get to the targeted weeds. If the aimed for weeds are eaten first, a continual low intensity effort may prove best.

The time of year (season interval) can be important. A weed, not normally consumed, may be preferred when young and tender. At the end of a dry season, when the most savory herbaceous plants are not procurable, less desirable forage will be sought.

More can be done to make grazing an effective tool. This can involve changing the weed mix through management (e.g., the ecosystem design or the use of fire) or through an appropriate weeding strategy (e.g., removing only the taller woody species).

Although the study is lacking and few employ animal species to full advantage, grazing has proved a befitting tool. Valderrábano and Torrono (2000) found goats capable of controlling specific scrub species in pine stands. Revkin (2000) noted that goats can also prune branches if this is the best available food and the plant is at the right height.

As with fire, the rule is for caution, foresight and understanding. Some grazing strategies can be ecological benign, others have greater impact. The differences can be pronounced. Benson (1977) observed that in areas grazed by sheep, the cactus species *Opuntia basilaris* var. *treleasei* is found, but in areas grazed by cattle, this species is absent. Lacking foresight, most of the danger occurs where animals roam widely, are in great numbers, and their labor is not focused.

PRUNING

The higher wood value of clear, knot-free logs favors stem pruning. Although mostly thought of as a management activity accomplished manually, natural forces do contribute.

Hand Pruning

Although a number of pruning techniques exist, only one is widely used in forestry. The five pruning methods are: (1) coppicing, (2) pollarding, (3) branch pruning, (4) lopping, and (5) stem pruning. The dominant silvicultural technique is stem pruning.

Briefly, coppicing cuts a tree or shrub to ground level where the tree resprouts from the stump. This has use in agroforestry to increase the number of stems for grazing. In silviculture, application is more limited, serving more as a regeneration technique (see planting methods, this chapter).

Pollarding cuts tree stems at a set height (2-3 m). This has a purpose in agroforestry as a means to briefly eliminate a plant as competitive force above seasonal crops and to protect regrowth from grazing.

Branch pruning is where branches that originate from the central stem are not pruned, but secondary branches are cut from the main branches (see Photo 1.2c). This is an agroforestry technique to open a canopy for increased light penetration and has little direct silvicultural application.

Lopping is where the outer branches of tree are cut. The purpose can be to reduce canopy spread, gather more horizontal light, or to increase fruit set. Early foresters saw some value in lopping to increase the amount of heartwood in stems, thereby increasing the value of a future harvest (Charpentier, 1902).

Stem pruning remains the dominant forestry method where those branches growing off the main trunk are cut, leaving a smaller, higher-up canopy. Well implemented and subsequently well managed, this produces clear, knot-free, higher-valued logs.

Natural

For natural pruning, two ecological forces interact to make this a success (1) the natural ability of a tree species to shed dead branches quickly after they die and (2) a stand design which, through internal shading, stimulates the lower branches to die. The first of these is a genetic attribute and a DPC, the second is a function of below-canopy ecosystem light.

For light demanding trees, this occurs at higher light levels, for shade tolerant species, this occurs only within a well shaded ecosystem. Coniferous species have advantages as, once the branches die, these are not replaced. For many or most broadleaf species, branches may be replaced if the amount of understory light increases.

Figure 6.1 shows a tricultural formulation and the sequence that leads to ecologically imposed pruning. This starts with primary species and slightly slower growing, dense-canopied, guide species. The guide species forces the primary trees into a welcomed, straight, branch-free stem form. Once the canopy closes, the secondary species, not liking the stand-internal light conditions, dies. Multiple secondary species produce similar results (as described under notable variations, Chapter 10).

These dynamics, in their most beneficial form, need to be better understood before these can be truly effective. Groninger et al. (1997) found that, with a mixed black locust-pine plantation, self-pruning occurred with the black locust, but was not pronounced in the pine. The problem was reported to lie with intra-species height differences and with a wider-than-normal pine spacing that allowed excess below-canopy light.

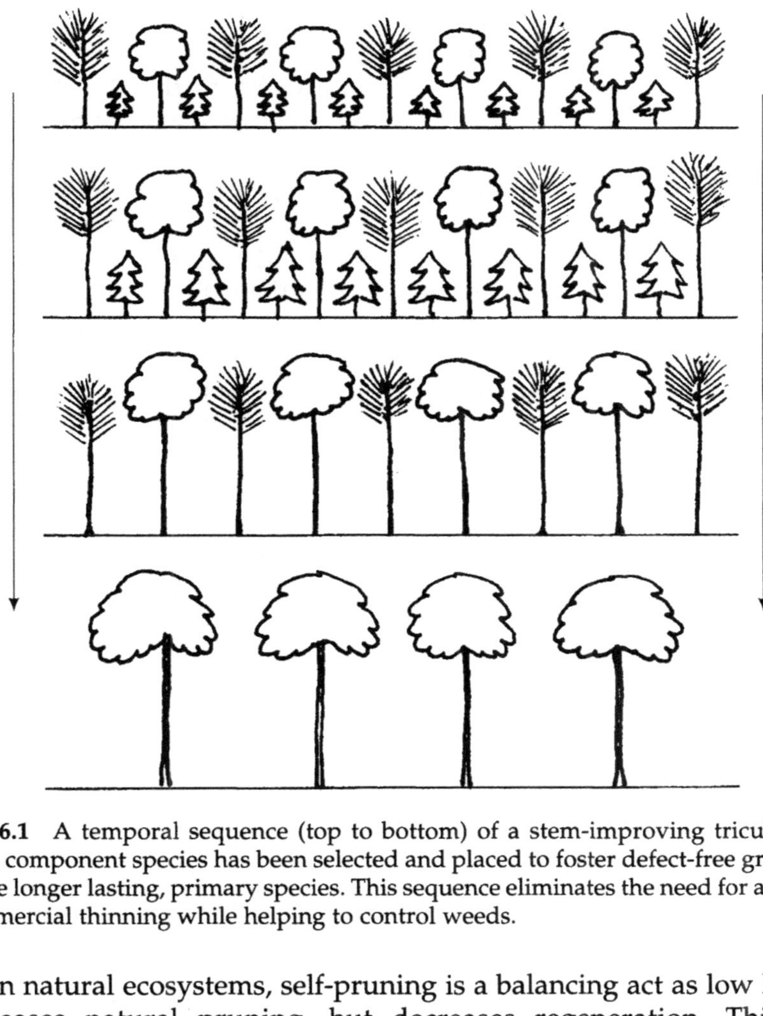

Fig. 6.1 A temporal sequence (top to bottom) of a stem-improving triculture. Each component species has been selected and placed to foster defect-free growth in the longer lasting, primary species. This sequence eliminates the need for a non-commercial thinning while helping to control weeds.

In natural ecosystems, self-pruning is a balancing act as low light increases natural pruning, but decreases regeneration. This is overcome through progressively more shade-tolerant species (as occurs in natural successions), leading to a self-pruned climax stand.

THINNING-HARVESTING

The thinning regime is part of maintaining an economically optimum density. Lesser densities waste essential resources (and may reduce stem quality through excessive branching), higher-than-optimal densities crowd trees, reducing growth rates (but this can help suppress weeds). Somewhere between, along an ever-changing timeline, lies the optimal density.

As mentioned in earlier chapters, when to thin the trees is a economic compromise, where the cost is weighed against the benefits of increased growth and better stem form. Other management decisions, such as fire danger and weed growth, also figure in this calculation.

Non-commercial thinnings are those undertaken at a financial loss where it costs more to harvest the tree than the revenue received. Some may non-commercially thin with the hope of recouping the cost with a higher valued, ending harvest. In other situations, excess, growth-reducing densities may be tolerated until the to-be-removed trees are larger and the stand can be profitably thinned.

Agroecology, through well-intended mix-species plantations, helps in this regard with gains coming along a number of fronts (weed control, better stem form, etc.). As shown in Figures 6.1 and 10.8, a guide species can both improve stem form and eliminate the early thinning. If done well, the first manual thinning is also the first in a profitable series of commercial harvests.

CHAPTER 7

Risk Containment

Forest ecosystems are subject to a range of threats. Strong winds, tornadoes, hurricanes and storm-induced micro-bursts can break stems and/or topple trees. Fire danger can be ever-present and destructive; insects and disease, a constant menace; drought, or the opposite, water inundation can have negative consequences; and hungry animals can consume large amounts of biomass.

Productive forest lands support more natural flora and fauna than their agricultural counterparts. This means a greater imperative for preserving the character of the land and for utilizing ecologically benign inputs.

This goes to sustainability, avoiding harmful inputs and/or changes that can affect the ecology and/or productive capacity of ecosystems. Adverse events include severe fire or extensive blowdown that harm yields or large doses of insecticides, fire retardant or other chemicals may impair water, wildlife or other forest resources.

Rather than subject ecosystems to events that may prove ecologically or productively non-compatible, an agroecological approach favors in-place, risk-reducing countermeasures. If an outbreak or an unwanted event were to befall, remedial actions taken should be as a progression, starting with the environmentally benign, carrying on into those deemed more severe. Discussed below are those countermeasures that can be permanently stationed or temporally initiated, upon need, into susceptible stands.

WATER

The effects of water, as with flooding, can be immediate and severe or, as with drought, long and lingering. Although these are opposing forces, these share some common countermeasures.

Infiltration reduces water runoff and increases groundwater. Groundwater is far less destructive than surface water runoff. Better infiltration can also alleviate drought. Well-stocked groundwater resources, replenished after brief rains, keep plants growing and healthy long after precipitation has ceased. In contrast, surface runoff is a very temporary affair.

Control

The danger begins when rain falls on unprotected soil, dislodging particles. This continues when water, running across soil surface, carries away these loose particles. Adding to the damage is the leaching of essential mineral resources.

Slowing surface water movement helps, infiltration of surface water into subsurface layers is better. Subsurface movement traps particles and slows movement so that water is available to plants for a longer period. Additionally, waterborne minerals are more prone to capture by plants with a lengthened exposure.

Control relies upon two mechanisms: (1) a ground cover and/or (2) barriers. Both can slow water, allowing it to penetrate into the soil structure.

Cover

Protecting bare earth with organic matter is cover control. This can be as living plants; cover crops serve well as do thick shrubs. A humus layer of dead organic material (e.g., fallen leaves, small branches, decaying stems, etc.) suits this purpose.

Without a humus layer, living trees overlay the soil, providing good protection as long as the lower leaves or needles are close to the surface. Effectiveness is reduced if the canopy is too far above the ground.

If distant from the ground, the barrier effect of leaves and branches on surface flow is lost. Additionally, the impact of raindrops striking the surface may not be lessened. Above 10 m, the impact force of water dripping off leaves is the same as or greater than unimpeded rain. Compounding the problem, the on-soil impact for leaf-origin droplets, after having consolidated on leaf surfaces, may be greater.

A well-formed humus layer, in addition to raindrop impact protection, can hold considerable amounts of water. This subtracts from surface flow potential while keeping moisture on site and available to plants.

Photo 7.1 Risk reduction through water management where (a) logging slash forms a soil protecting ground cover and (b) an infiltration ditch positioned to reduce erosion and improve in-soil moisture.

The presence of deep, well-formed humus provides good erosion control, but any soil cover will serve if more than 70% of the surface is uniformly overlaid. In older monoculture plantations, the humus layer and the surface roots, rather than the aloof canopy, are the principal ground cover.

The amount and type of organic materials produced figure into ecosystem design calculations. Larger leaves provide more surface soil protection than do small leaves, slower decaying leaves provide more protection than those that decompose quickly. Besides being DPCs for water management, leaf characteristics can also be part of a weed control strategy (Chapter 6).

Barrier

In silviculture, often with overriding cost considerations, a cover approach to erosion control and infiltration can be more significant than barriers. Despite this, barriers are an alternative, e.g., where tree establishment is fraught with difficulties (see Chapter 6) or where periodic droughts threaten in-place trees.

Barriers can be ditches, hedgerows, uninterrupted mounds, strips of live vegetation or any combination of these. The purpose of any barrier structure is to keep water in place, giving time for the moisture to soak into the soil. Normally, barriers are perpendicular to water movement and contour hillsides, although exceptions to this general rule exist. The exceptions include riparian zones that follow watercourses and semi-contour ditches or bunds (Wojtkowski, 2004).

A number of factors determine spacing and actual design of structures. There is a tradeoff between frequency and size; a few small structures or widely spaced larger barriers define the alternatives. In practice, effectiveness is gauged by observing leakage over or through a barrier after an above-average rainfall.

Barriers can be designed into ecosystems through spatial pattern, be provided by an auxiliary structure (as with riparian buffers) or be landscape wide. As with most water control measures (barrier or cover), the usefulness is determined by the overall impact across a landscape.

FIRE

Fire is as a useful forestry tool (Chapter 6), a factor in the evolution of ecosystems (Chapter 14), as a means to counter insects and diseases (later in this chapter) or a productivity threat. Threat containment can lie with nearby fire crews and the assumption that they will make short work of any immediate flare-up. The actual work undertaken by fire crews is outside the scope of this text, however, in-situ preventative measures are part of the ecosystem and/or landscape design. As with water management, the mechanisms that hinder fire movement are cover or barrier based.

Barrier

A fire barrier is a key element in fire management, only a small number of regions have the rainfall patterns and moisture retention where fire barriers are not essential. Where needed, natural and manmade features of the landscape serve a barrier function, steams and roads are the most common. It should be noted that a road can also be a source of danger (burning object thrown from vehicles), often mandating along-road barrier protection.

Dedicated firebreaks are usually open strips between forest stands or plantation blocks. These come in varying forms (modified from Bellefontaine et al., 2000):

(a) cleared and bare,
(b) cultivated,
(c) grazed,
(d) containing a fire-resistant species, and
(e) employing fire transmitting vegetation.

The cleared barrier is the most common type. These can be part of an access or logging road network or bareground managed specifically for fire protection. The latter requires maintenance, provides no ecological gains (such as those associated with biodiversity) to adjacent ecosystems, and does subtract land and economic potential. The width is such that a canopy-level fire has difficulty crossing.

Cultivated or grazed strips are possible as long as these do not encourage fire. The advantage lies in their economic activity (which allows for wider strips) and in the ecological gains (such as those described under insect and disease management, this chapter). These strips are economically attractive if they provide valuable forage or crops. Climate permitting, crops will be present during a rainy season when fire is not a threat, missing when fire danger is highest. Tuber crops (potatoes, sugar beets, etc.) serve well. Upon harvest, these leave the ground bare.

Other plants can address specific requirements. Among the possibilities are strips planted with a fire-resistance species. These species not only survive fire, but help suppress minor burns. A few have been identified, e.g., many species of cacti are fire resistant as are some agricultural crops.

The final barrier type allows fire to cross, but in a controlled manner. This is employed where fire is a management tool. The only proviso is that, in crossing, flames must not reach high enough to spread to the tree canopies. Grazing can aid in this.

Cover

Fires seldom start in tree canopies. The source and main point of control is the ground cover. The type and management of the ground cover can impede spread or decrease severity.

Fire resistance (those species that do not promote spread) can be a DPC of a ground cover crop. Lacking this, grazing or even burning at the ground level under controlled conditions can prevent, by maintaining a low-level fire, far more destructive and undesirable crown fires. The guidelines for fire in fire prevention situations are the some as those for weed control (Chapter 6). The two, grazing and fire, in tandem can be a powerful tool, employed both to control uninvited vegetation and unwanted fire.

WIND

As a threat, wind is always present. A constant breeze can be a negative influence, the first by robbing plants and soil of moisture and through less noticed disservices such as leave and branch rubbing. The latter saps substantial energy and productive potential from growing plants (Brenner, 1996). Stronger winds break branches and stems and topple trees.

Counter mechanisms can be tree-based, stand internal or landscape wide. DPCs include strong stems, well-anchored roots, drought resistance and a low wind-drag canopy. Stand design details lie with the tree spacing or the planting of a secondary species to resist wind force or to interlock and help hold roots within the ground.

Landscape mechanisms are the placement of shelterbelts, windbreaks and other protective ecosystems.

LARGE HERBIVORES

There are a wide range of conservation-agreeable options. The more benign countermeasures include fences and repellent fauna. Repellent fauna can include ferocious dogs to chase away smaller grazers and, against the largest, the elephant, beehives placed in trees seems to work (Anon, 2002).

Where fences and the like do not work or are costly, hunting or trapping can go far in eliminating a threat (Giesser and Reyer, 2004). When conservation is paramount, wild herbivores, large and small, can be treated as a resource to be sustainability-harvested. Semi-husbandry, looking at wild animals much the same as domestic stock, is another option (see Chapter 15).

INSECTS AND DISEASES

The control of herbivore and wood-boring insects and plant diseases is a must. The attacking organisms are many and varied and require continual monitoring to assess the degree of risk. This often relies upon some threshold value, beyond which population dynamics and the destruction caused will make the risk unacceptable.

IPM (Integrated Pest Management) is often employed to monitor the level threat and trigger some counter-measure. Although not often purpose-defined (Walter, 2003), one view of IPM is to invoke a severe and immediate response that eliminates the risk. The common response is in the form of chemical attack with varying side effects and ecological risks. The advantage of IPM is in applying lesser amounts of toxic chemicals, by using only when needed, and by localizing the negative ecological side effects.

Agroecology offers a wide array of options, chemical application being a last resort measure. Under these circumstances, IPM is used, both to monitor the effectiveness in-place measures and to trigger the steps in a progression of responses.

Basic Countermeasures

The countermeasures range from landscape wide to local application and can be designed for a broad range of pests or for specific scourges.

Most apply to insect dynamics, a few have application against disease-spread.

(1) Healthy populations where plants under less stress resist-attacking organisms better. Less crowding and greater inter-species facilitation can yield hearty, resistant trees. As a management tool, density modification with ample resources to produce sound trees has been employed effectively against pine beetles (Waring and Pitman, 1985).

Along these same lines, plants can be made more unpalatable through immunization. This is accomplished by spraying an area with chemicals that induce natural resistance (Day, 2001; Ruc, 1990). This may be more of a measure when outbreaks loom.

(2) Rotations to rid the soil of unwanted residents are a mainstay in agriculture. Forestry, because of the long rotations, cannot take full advantage of this affect. Usage is mainly confined to agroforestry and the possibilities for periodically changing the understory vegetation to counter a specific pest.

(3) Predator insects have a dietary preference for herbivore insects. The choice is large and, in use, so should be their appetite. The operational strategy can be (a) for a general habitat, attracting a diversity in predator insects, or (b) to focus on a few specific predators; ants and spiders are commonly sought.

The use of whatever predator insects nature offers is the easiest approach. The key is in maintaining sufficient predator populations through habitat and biodiversity. A thriving understory, weeds included, will draw and hold large populations of beneficial bugs (Altieri and Nicholls, 2004).

Specific predators are attracted by specific plant species. Ants are lured with flowering plants in the understory or secondary tree species with attractant properties as a DPC. This instrument may involve less predation, more molestation (chasing unwanted insects away). Whatever the case, ant-infested trees can experience greater growth (de la Fuente and Marquis, 1999) but, as a strategy, specific predator ants may be preferred over general populations. In rainforests, some ant species devour large amounts of biomass (Davidson et al., 2003).

Spiders have a greater appetite for other insects (Marc et al., 1999). These also have a lighter touch, i.e., in not eating plants.

Predictably, and with scant notice, predator-prey relationships function below ground. Although far less understood, these unseen interactions can positively influence ecosystem dynamics (Preisser, 2003).

(4) Fire is a common overall tool but, against insects, the gains are less understood and mixed results have been reported. Agricultural studies support fire as an insect control method (Stringer and Alverson, 1994) and this extends partially to silviculture. Anderson et al. (1987) found that fire reduced the intensity and duration, but not the frequency, of spruce budworm attack in the western North America.

Other studies have found the reverse. In western North America, Santora et al. (2001) found fire increased the number of beetles and these were more successful in attacking fire-damaged trees. Wingfield and Swart (1994) observed that burning, in this case, ground slash, contributed to the spread of a root disease.

Clearly, this is a insect- or disease-specific tool. Timing and intensity being part of any in-field strategy. At this point, these are not well understood.

(5) Encouraged or introduced large predators are an insect-control tool and part of an overall reduction strategy. Bats and birds eat insects and can be favored by local habitat, especially where habitats are being purposely maintained. These can be through bird or bat houses, favorable tree species, stands, and a host of other population encouraging additions.

Woodpeckers, for example, seem to damage trees but, in actuality, help more than they harm (MacLellan, 1970). With a diet primarily of insects, these birds nest in dead trees and can be encouraged to stay by offering plenty of standing dead timber.

(6) Attractant crops (to harbor predator insects) include those flowering species that lure ants and similar insects. This attribute can be inclusive in the DPCs of a covercrop, secondary or neighboring species (one or more).

General biodiversity also works well. The usefulness of weeds in this role has been discussed (Altieri, 1994). Bragança et al. (1998) noted that pests in a eucalyptus plantation were curbed by remnants of nearby native vegetation and the populations of predator insects contained.

(7) In-situ passive repellent vegetation includes large groups of plants that repel insects. Many are known and these can be a facilitative addition, purposely placed to fill this role.

This can occur as an above ground or below ground affect. The latter works against nematodes and other root-attacking organisms. Some plants are more potent that others and, again, this can be a DPC of one or more secondary species.

(8) In-situ repellent plants with traumatic release are where repulsive plants are made more effective if the volatile repellent chemicals are released into the air. This is done by damaging the leaves to release key compounds. As a result of the effort involved, this is more of a post-outbreak measure.

(9) Biodiverse barriers that impede the movement of damaging insects. The mechanism is a general habitat (see countermeasure 3 above) that shelters and encourages a mix of predator insects. In use, Cappuccino et al. (1998) found that patches of biodiversity thwart spruce budworm.

(10) Borders or barriers of specific plants to interfere with insect and animal movement. For this purpose, species that predator insects find to their liking or repellent plants serve well. As a living fence, these can keep out unwanted animals while answering a need to contain insect spread. These plants can repellent insects, to attract specific predators (e.g., ants), or be a mix of mechanisms. Vegetative fire strips can also serve as an insect, animal or disease-control border.

An example of a barrier tree is the African species *Acacia drepanolobium*. This tree has symbiotic relationship with different ant species. Some viciously defend the tree from herbivore insects and animals while removing disease threats (Stanton and Young, 1999). Ants, coupled with large thorns, make this a good candidate for an herbivore insect-destructive animal barrier (Stapley, 1998). Thorns or spines are also a deterrent for small climbing mammals (Cooper and Ginnett, 1998).

(11) Trap crops are those that specific herbivore insects find to their liking, more so than the adjacent crops. These collect greater populations of herbivore insects. Alone, trap crops are only a temporary measure. Long-term influence is gained where there are part of

(a) a chemical application strategy where only the insect-laden plants are sprayed; or

(b) a predator strategy where trap crops are purposely placed near high populations of predator insects.

(12) Micro-climatic conditions can discourage disease organisms (Koech and Whitbread, 2000) and insects (Schroth et al., 2000). The mechanisms include a less humid micro-climate to halt a disease or an internal system ambiance favorable for the spread of an insect attacking pathogen.

One application may be with patch size. This variable has been shown to alter the populations of herbivore insects (Bach, 1988). Generally, the micro-climate approach lacks study but, if developed, could find use against specific diseases and insects.

(13) Physical insect traps do catch insect species. A number of types are possible; those containing a pheromone bait and an insecticide (lure and kill), cloth or paper strips with a chemical attractant and a targeted pathogen (lure and infect), and an attractant to get the targeted insect to enter an imprisoning contrivance (lure and contain) (Suckling and Karg, 1999).

Less developed are traps specifically designed to snare forest-based herbivores or wood-eating insects. Again, this is a species-specific approach, one to be employed with outbreaks.

(14) Cut-and-carry repellent plants are those species that contain repellent chemicals. Instead of growing in place, these are cut and carried to where needed. These have limited forestry application, confined to small-scale outbreaks situations.

(15) Chemicals, benign and/or spot applied, which are designed to mitigate environmental damage. These can focus on a specific insect species and/or be spot applied. There are some home remedies that are not environmentally damaging and do well against specific pests. As an example, a meal made from milkweed seeds destroys nematodes and armyworms (MacKenze, 1999).

(16) Chemicals, harsh and/or broadly used, which aim to stop an insect or disease that threatens to wipeout a plantation crop. This is a last-resort strategy due to the cost, application difficulties when large trees are involved, and the environmental harm that can be inflicted.

Immediate results notwithstanding, the elimination of the herbivore insects is far from assured. There are cases where widespread spraying had the predicted result, but because predator insects were also destroyed, the situation was worse after the chemicals subsided and the targeted insect returned in greater numbers (van der Valk et al., 1999).

Two-pronged Defenses

There are situations where, instead of one defense, two measures working in tandem can accomplish more. Dedek et al. (1989) found benefits with an insect-attracting pheromone to lure bark beetles to an insecticide paste. This lure and kill example used an insect-specific pheromone in concert with a broad-spectrum, general-use, but spot applied, insecticide.

In this case of paired countermeasures, one is broad based and the other insect specific. Other applications are formulated around generalist and specialist approaches (Vandermeer, 1995) and, as with a two-pronged or dual defense, these can be coupled countermeasures.

These function in a number of ways. As in the above, one can help the other, or alternatively, if a generalist, and environmentally benign, strategy fails or is overwhelmed in the face of an onslaught, a landuser can resort to specialist control. More possibilities exist as where two generalist measures can work together as can two specialist measures.

Generalist

The generalist measures work, not against one pestilence, but keeping a range of insect pests or disease organisms under control. These can be plants and micro-ecosystems that retain generalist predator insects, including favorable habitats for ants and spiders.

Plants that repel all or most herbivore insects can also be part of a generalist strategy. The lists of these species are long and varied and can be looked at as a DPC.

Specialist

Where there is the danger that an outbreak of a specific species of insect or disease will occur. Specialist countermeasures focus on this immediate threat. Counter methods can be specific predator insects, e.g., ladybugs attacking aphids, wasps against beetles, etc. Specific chemicals can be employed, e.g., boring insects might be opposed with stem-applied chemicals (as mentioned).

Tracks

A more advanced version of the two-pronged defense is of a series of defensive tracks, each complementing the other, each adding overlapping layers of defense. If one defensive measure proves weak, another is in-place to accomplish the task.

Among the reasons for the tracks are that one insect-eating species should not devour other insect-eating species or that repellent plants should not drive away only the insect-devouring species. Well thoughtout and implemented tracks insure a degree of inter-defense harmony.

These might begin with the more benign, generalist approaches and end with something more environmentally harsh. Where management inputs are needed, IPM monitoring is the trigger mechanism for shifting from one defensive to another and ensuring that these are not overused, e.g., in spanning a greater area than necessary.

A Triangle Defense

As tracks proliferate, the option or strategies presented assume a triangle form. Diagrammatically, this is shown in Figure 7.1. Each block is a defensive countermeasure and each vertical column is a series-integrated track with inter-defense harmony.

The countermeasures range from the more numerous, integrated, and more benign (Figure 7.1, left side vertical arrows) to the simpler, more severe, and less selective (right side arrows). A single defense strategy (the single block on the right), hopefully one not employed, is inclined to be a severe chemical control.

The upper levels left side are the generalist, in-place countermeasures. The middle more transient, the lower specialist and short-lived. The in-place counter measures are generally landscape wide, control a number of insects species (and some diseases), and require little monitoring. The lower levels of a multiple, stacked defense (left side, Figure 7.1) are more specific with regard to the types of insect and disease organisms regulated, more labor intensive and require closer supervision.

Where the stacked defenses are fewer (right side, Figure 7.1), these tend to be generalist and severe. Suggested examples of individual vertical tracks (modified from Wojtkowski, 2003) are (as left side measures):

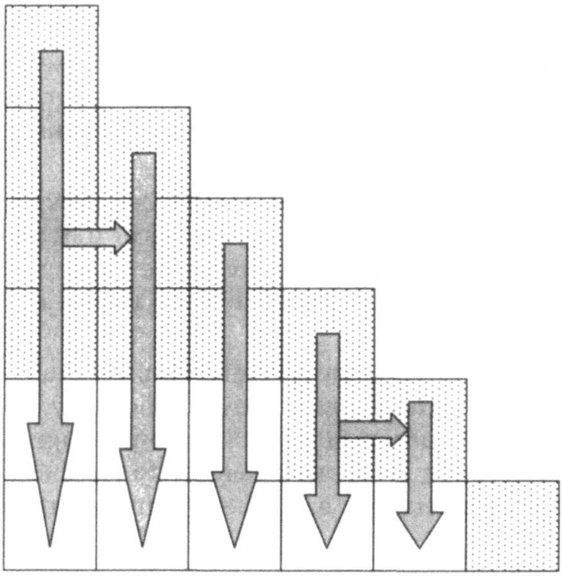

Fig. 7.1 Integrated risk abatement represented as a stacked countermeasure, each box being one remedy. The darker countermeasures are generalist, the others specialist. The more numerous, less severe, stacked countermeasures are on the left, the more intrusive measures on the right. The vertical arrows represent coordinated control strategies with crossover (horizontal arrows) possibilities.

- a landscape conducive to predator fauna (e.g., insect-eating birds)
- movement barriers
- predator-prey hedge rows or grass strips
- cover crops or mulch to promote predator movement
- repellent plants located so as to drive insects toward hedges
- cut-and-carry strips for outbreaks
- introduced predator insects

some mid-column defenses are:

- spread prevention barriers
- bird and/or bat houses to retain unique predator fauna
- trap crops to lure and concentrate a specific herbivore insect
- host plants to retain predator insects
- introduced predator insects to supplement natural populations

a second to the last column block series can include:

- interspersed trap crops
- the spot application of insecticides

and as a last column (furthest right) defense:

- a broad spectrum insecticide

Other options lie along the diagonals, shifting from one track to another (horizontal arrows, Figure 7.1). This might include (with the above hypothetical orderings) as (a) a landscape conducive to predator fauna, (b) spread prevention barriers, (c) interspersed trap crops.

At present, these are mostly abstractions. Before individual tracks and a full triangle of options accrue, more needs to be done in understanding the threats, the countermeasures, and how these are ordered and can be made to work in harmony.

INCLUSIVE APPROACHES

Outlined in this chapter are various threat countermeasures. The option favored here is to pre-place defenses so that threats are lessened. There may be disadvantages, e.g., a second species, added to drive away harmful insects, can reduce growth rates in the primary species or loss in growth can occur with controlled burns. There can also be overall gains where a well chosen facilitative species augments tree growth while guarding against some unwanted happenings.

The control options extend to multi-stand level where risk reduction is based upon biodiversity, not only in primary species, but in accompanying or neighboring vegetation. Stands susceptible to one threat may be defended against this threat while a nearby stand, more susceptible to another event, may be designed more to withstand this occurrence. Fire-resistant stands interspersed with fire susceptible ecosystems are examples of this approach.

Topography and relative location also figure in establishing priorities and in positioning countermeasures. Without going into the details of the landscape as a means to counter risk, it is easy to see that options extend well outside the providence of a single stand. Most often a combination of options coexist.

COMPROMISES

The matching of a stand design against risk influences may be the easy part. Defenses notwithstanding, it is not always possible to guard against everything while gauging the types, magnitudes and frequencies of threats is far from a precise undertaking.

An unanticipated attack by a previously unseen insect species, hitherto not considered a danger, is a possibility. Even if this were to occur, a basic design, one with generalist protections, should offer some immunity. Insects are classic case where in-place countermeasures should help guard against the unanticipated.

After all is said and done, events do not always go as planned and stands cannot be completely shielded. An incident of great magnitude (e.g., an insect plague or very high winds) can overwhelm a stand once thought to be impervious.

In an agroecologically perfect world, knowing that fire will not go far in a fire-resistant or fire-maintained (frequently burned) stand, it should be possible to let it burn. Fire resistant species and/or in-place measures allow this to happen.

Take the case of improved varieties of radiata pine, borrowed from wetter New Zealand, that are planted in fire-prone Australia and Chile. Countermeasures such as fire breaks and fire crews are in place, but only offer limited protection. More proactive in- and ex-stand countermeasures could certainly improve the situation. Fire-resistant species, e.g., loblolly pine, are available, but not utilized. This may be because the degree of domestication, in terms of local growth rates, are less.

Risk and protection analysis does not always lead to a satisfying outcome (Rabin and Thaler, 2001). How much risk is accepted lies in the eye of the practitioner.

Small flare-ups and the occasional large conflagration represent the negative side of the radiata pine compromise. The reasons are economic. The losses in privately owned trees are an accepted, but undefined and non-budgeted cost of doing business.

What can tip the balance to increased danger is that fire crews are often a governmental responsibility and the environmental toll that fire inflicts on nature is not a business expense. On the other side of the financial equation, there are costs in reformulating a fire-susceptible monoculture to something less vulnerable.

Although much can be done in establishing systems of defense, risk assessment is less of a definitive measure and more an arbitrary stipulation on the part of individual landusers. Still, preplaning along agroecological lines should relegate these to what they are; highly unanticipated or unforeseeable happenings.

CHAPTER 8

Monoculture

The monoculture, a single species or plant type in a given area, is the simplest of agroecosystems. As all species are niche-identical or niche-similar, these are, in ecological terms, a constrained fundamental niche. Monocultures are species-governed where the plants compete spatially for a nearly identical intake of essential resources and thwart adversity mostly through desirable plant characteristics (DPCs).

Being uncomplicated, they are a favored agriculture and forestry practice. The lack of bio-complexity notwithstanding, a number of variations exist, each with sound economic and even ecological rationale.

REASONS FOR USE

Simplicity drives research and research drives monocultural furtherance. The absence of underlying agrobionomic theory and the ability to direct all essential and labor resources to a single output is also attractive. With proven outputs, ease of monitoring and economic simplicity, these systems are favored by industry and lending institutions (Godoy and Bennett, 1991).

Popularity may also arise from a cultural bias. Many developed regions have a cultural proclivity for simplicity and order. The ecological advantages and resource-use efficiencies of bio-complexity seldom trump this.

Monocultures are not without parallel in nature. Pines, redwoods and eucalyptus can exist in fairly pure stands. Pines discourage other vegetation through soil acidity, redwood through alkalinity and eucalyptus is thought to exclude other plants using water as the limiting essential resource. Outside support, in the form of fire, often leaves pure stands of fire-resistant trees.

In practice, monocultures mimic the role that species play in early succession, where fast growth and an ability to overcome, resist, and/or suppress weeds carries across to plantation plantings. Species more agreeable to intercropping, i.e., those often associated with latter stage or climax forests, are found less in plantation monocultures.

TYPES

A common definition of monoculture is, by intent, the cultivation of a single species. Except in the most stringent interpretations, this definition allows considerable in-practice flexibility. Thus, there are a number of monoculture variations,

(1) pure (niche-identical);
(2) niche-variable;
   (a) varietal,
   (b) genus,
(3) uneven age;
   (a) coppice with standards,
   (b) uneven plantings,
(4) non-interference.

These are not exclusive; combinations are possible. A non-interference monoculture, where other vegetation is permitted or tolerated, can combine with pure, niche-variable, or uneven-aged versions. The pure or niche-variables types can also be uneven aged.

Pure

A pure monoculture is the truest form, having only a single, niche-identical plant species. This defines a clonal relationship where the plants are genetically alike.

The pure clonal monoculture is uncommon, but not unknown in nature. Genetic likeness may be found where a stand grows from root or stump sprouts, e.g., bamboo groves.

In silviculture, a common commercial strategy is to pick a superior variety, one that possesses the DPCs to resist all anticipated problems, clone, and plant these in large numbers. The longer timeframe and the tendency to have large areas of a single clone runs counter to what predominantly occurs in nature.

In doing this, risk can be heightened. An attack by an organism that has evolved to overcome in-plant resistance and take advantage of this abundant food (tree) source is one possibility. Another is the arrival of some foreign pest that is not deterred by the current DPCs.

Niche-variable

The truest of monocultures is one, populated by clones. Other types have similar species, but encourage some genetic difference. This is the norm in seed-generated natural monocultures that incorporate minor amounts of genetic diversity.

Some tree species, particularly those with wide geographic ranges, can offer substantial diversity across the range, (Rehfeldt, 1978). How valuable this is and how it is best engaged is an open question. For example, is it better to find trees well suited to local conditions or to seek and mix trees from the fringes of the range to better tolerate climatic and other variations? Without clear guidelines, this is an inexact science.

Varietal

The varietal monoculture exploits the genetic advantages of varieties in combating natural stresses. Disease and insect infestation are principal among these. For this, varieties of a single species are interplanted with the idea that, in the event of infestation or natural calamity, the resistance variety will survive, others will be stunted or lost. The difference can be expressed through the DPCs.

The proliferation of poplar varieties in northern Europe can champion this approach with trees being selected for subtle properties. Growth-rate differences, if any, are handled as in a three-plus polyculture (see Chapter 10).

However, documented silvicultural studies are lacking. In agronomy, successful applications show a clear varietal ability to deter disease spread. In one case, rice yields increased 200% over what could be expected if the prevailing diseases did not encounter a genetic barrier of varietal origin (Yoon, 2000).

Genus

Varietial monoculture has recorded ecological advantages, but the better silvicultural option may lie with a species-rich monoculture. Wood properties tend to be similar within species of one genus and plantations can be formulated within a genus class.

Genus monocultures are best when a number of species of the same genus cohabit a region. In Japan, the choice for a plantation of native pines may be a mix of *Pinus densiflora*, *P. thunbergiana*, and *P. parviflora*. For Spain, the selection is better and includes *Pinus pinea*, *P. nigra*, *P. mugo*, *P. pinaster*, and *P. sylvestris*. In contrast, Mexico is flush in pine species, with over 20 local species available for selected sites.

Oaks, being a genus with numerous species, offer similar opportunities. A mix that might be contemplated for eastern North America may include, depending on site suitability, two or more of the following oaks: northern red (*Quercus ruba*), scarlet (*Q. coccinea*), pin (*Q. palustris*), black (*Q. velutina*), water (*Q. nigra*), laurel (*Q. laurifolia*), white (*Q. Alba*), bur (*Q. macrocarpa*), and/or chestnut (*Q. prinus*). If the market demands wood of a uniform color, a planting can mix red or white oaks. Alternatively, a plantation could also be formulated outside of a color grouping; wood color being an end-use perspective.

In the above possibilities, it is conceivable that the adding of exotic pines or oaks would not affect the output value. Mingling native and exotic trees of the same species may ecologically mask the introduced species, allowing it to coexist within an ecosystem where it is less foreign (assuming that exotic pests are not also introduced).

It is conceivable that members of a family, e.g., Pinaceae (pines) and Fagaceae (oaks), rather than the genus be utilized. This approach can lose the advantage of a like output; the product gain being no different than that of a true multi-species polyculture.

With a genus mix and its built-in genetic diversity, broad-spectrum resistance to destructive organisms and other ecological gains should accrue (better yields from better resource partitioning). Complementarity is desired and, if found, is exploited as with other multi-species associations.

Since these are separate species, the DPCs will differ and this should be exploited to full advantage. Bicultures and three-plus polycultures (Chapters 9 and 10) are often better models for explaining composition, layout and other application nuances. Although genus-based monocultures can offer ecological and commercial potential, examples are extremely rare.

Uneven Age Plantations

Many view monoculture as containing plants of a similar age. Although not a requirement, it remains a common perception. Shedding this notion adds to the list of possible variations.

In newly-planted areas, there are untapped essential resources. The same holds true in a well-thinned final stage plantation. It is possible to place newly-planted trees in the open spaces between the mature trees. As a benefit, young trees are introduced into a functioning ecosystem that can protect and increase survival. A second gain comes with the temporal overlap and the shortened timeframe between harvests.

Some ecological and economic gains occur if not all the trees are cut in the first rotation. Having a few large logs provides increased value in future harvests. Older, taller plants within an otherwise juvenile stand can provide wind and frost protection. Likewise, a crowded understory produced by natural re-seeding is a form of ground cover. Being mostly undocumented, uneven-aged plantations are discussed in some detail under notable variations (this chapter).

Non-interference Monocultures

The idea that, as long as there is no substantial interference, other forms of vegetation can exist in what should be a pure stand, one still classed, through intent, as a monoculture. Non-interference is the situation in which:

(1) many mature forestry monocultures where herbaceous or woody, ground-level plants and a large assortment of in-canopy vegetation are found in various quantities;
(2) a newer monocultural plantation where extraneous vegetation exists before tree growth and subsequent suppression takes hold; and
(3) the post-clearcut replanting of natural forests where even aged, pure stands are planned.

Often trace species, these may or may not contribute to overall system dynamics. Lacking a visible or definitive measure, the ecological gains might not rise to the level of an ecological truth and, in the eye of the beholder, the system does not cross the classification line into the realm of an interacting polyculture.

In not interfering with growth or classification, these seemingly superfluous plants are ignored for economic, rather than ecological reasons, i.e., the cost of removal is greater than any lost growth. Nonetheless, weeds add biodiversity and confer ecological gains (Altieri, 1994). If of sufficient magnitude, a bottom-up ecosystem may evolve, conferring some of the ecological gains associated with ecosystem governance (enhanced nutrient cycling, control of herbivore insects, etc.). Photo 8.1 is of a non-interference monoculture having the type of low density biodiversity that would minimally influence tree growth and ecosystem classification.

Photo 8.1 A well-spaced pine plantation with a sparse, non-interference understory.

A cover crop in an otherwise monocultural planting does not constitute a non-interference monoculture. Technically, this is a biculture and these are discussed in this context (see Chapter 9).

DPCs

For monocultures, whatever the type, the preferred DPCs are:

(1) site compatibility,
(2) fast initial or overall growth,
(3) self pruning,
(4) good stem form (depends on end use),
(5) good wood quality (again depends upon end use),
(6) ability to resist insect pests and diseases,
(7) ability to resist or suppress competition,
(8) ease of establishment (either through coppice growth, seeds, etc.), and
(9) a tolerance to reasonable climatic variation.

There are also special need characteristics (e.g., that relate to site requirements or are desirable, but do not dominate or overly influence the selection). Where a number of species possess the above characteristics in more or less equal measure and a selection decision needs to be made, other characteristics are tie breakers. The expanded (desired, but less compulsory) DPCs are:

(10) fire resistance,

(11) ability to produce large amounts of soil protecting humus,

(12) large leaves (helps with weed control, erosion protection, and in holding in-soil moisture), and

(13) self thinning (a not-so-rare and obtainable quality where the slowing growing plants naturally die).

One DPC not listed is that of having a good understanding of the tree species in question. A lot of species have the potential to become plantation trees and are not employed simply because these have not been identified as such.

In practice, there are generally a few species that find service within a given region. In the temperate southern hemisphere, radiata pine is a leading plantation species; in the southeastern USA, loblolly pine fits well; and, in northern Europe, poplar hybrids are grown.

Some plantation species are a vestige of a past where the end use has changed and/or where rapid growth was less important. Oak was once needed to build ships and planted for this purpose. Teak plantations in tropical regions are a relic from an era where slow growth and long rotations were the accepted norm.

MANAGEMENT OPTIONS

To achieve the desired outcome, monocultures often require more in the way of active management inputs than with other practices. However, through species selection (DPCs) and system design, management needs can be reduced; as with the ability of the primary species to exclude unwanted vegetation.

Planting

The planting methods (Chapter 6) can accommodate different climatic and management situations. With the monoculture, the most common planting method may be through seedlings. These are employed where natural regeneration is not reliable and/or new genetic stock is being introduced. Striplings fill a similar need while offering some immunity from grazing animals.

The convenience of sprouts and coppice growth has not been overlooked. Many non-coniferous, commonly-found plantation species (e.g., poplar and eucalyptus), upon cutting, coppice profusely. Without the cost of replanting, this is attractive for re-establishment.

Large stem plantings also find application in intense, high-output, short-rotation situations. The gains that accrue can favor intense situations, e.g., wide spacing and accelerated growth eliminates the need for a non-commercial thinning. More potent is where the combination of a large stem planting and series overlap. This can reduce a rotation by one-third over systems planted with seedlings and lacking an temporal overlap.

Weeding

As so few plant niches are occupied, the proliferation of weeds is a major monocultural disadvantage. Species selection is a key variable, another is to combat unwanted growth through management inputs (manual or natural weeding) or employ the outside forces of nature (fire and/or grazing).

Manual

As many of the biodiversity options are missing, there may be a greater reliance on hand weeding. An entire area can be hand weeded at great cost. However, with forethought and planning, this can be avoided or the cost contained. Some of the applicable weeding techniques are:

(1) total removal;
(2) partial removal;
  (a) near the stem,
  (b) away from the stem,
(3) species specific; and
(4) natural exclusion.

Briefly, total removal destroys all weeds, either through hand, machine or herbicide use. There are a number of versions of partial weeding. These result non-interference monoculture or something more profound.

One version, near-stem weeding, removes only those weeds closest the stems and most likely to slow tree growth. Grazing and/or fire can take care of the remainder or, if inconsequential, these may be ignored. Again, the purpose is to reduce overall cost. Where grazing and/or fire are employed, there is a caveat for resistant trees.

In another version of near-stem weeding, the remaining weeds, those that no longer crowd the primary species, have a purpose (Taguder and Reyes, 1962). Under tropical, fast growth conditions, the result can be a vertical gap in which the primary species resides (as in Photo 12.3). This favors a plantation species that collects vertical, rather than horizontal, light. The tall, uncut weeds become a facilitative addition that guide and improve the stem form of the primary species.

This can eventually lead to a natural forest. Discussion along these lines continues under the heading of an in-plantation forest (Chapter 12). Another option, near-stem weeding as a step toward a successional plantation, is touched upon under noteable variations (this chapter).

Assuming that trees can tolerate or suppress nearby weeds without help, weeding can be done away from stems. This form of control may be employed in conjunction with burning or be motivated by the need to keep inadvertent fire from jumping from tall weeds, those less suppressed, onto tree canopies.

Although chemical weeding may be questioned on ecological grounds, total and partial weeding can be approached through herbicides. For selective weeding (below), planting killing chemicals may be too blunt a tool.

Species-specific weeding is where only those plants which most interfere with tree growth are removed. The others, those that are less competitive, are allowed to remain. In this non-interference monoculture, the weeds become, in essence, a form of cover crop or unintended agricultural addition.

The socioeconomic side to weeding removes those weeds without community value. Often unnoticed, weeds can be useful plants. These can be for food (berries, greens, etc.), animal forage, potherbs, medicinal herbs, and raw material (for basket weaving and the like) (Vieyra-Odilon and Vibrans, 2001; Turner, 1994). Leaving those with value may encourage gathering and local participation. It also must be remembered that weeds help in preventing erosion, in preserving soil quality and can be a component in an insect control strategy (Altieri, 1994).

Natural

The most potent ecological option, that of system-based or natural exclusion, employs, in monoculture, (a) trees that discourage weeds and/or (b) a high planting density such that weeds are shutout when the canopy closes. For the first, the attractiveness of species that self weed has resulted in this being a DPC. Pines through changes in soil pH, eucalyptus through the seizure of moisture, and others that grab, for a growth spurt, the lion share of available mineral resources are favored in this regard.

The double purpose of a high initial planting density, (1) for weed control and (2) the elimination of side branching and a more valuable stem, can make this option economically attractive, but only if planting costs are low and the initial thinning has commercial value. If not, it can be costly, although the expense may be partially recouped through a higher future stem value and reduced weed-on-tree competition. Trees with thick, light-blocking, canopies aid in this.

Fire

Fire is mainly applied where a plantation is fire-resistant. The southern (USA) pines are among many species with resistance and this DPC is exploited to reduce weeds.

Burning is done when the trees have reached an age and height where the canopies are out of harm. The process is generally undertaken during a wetter period and at intervals such that ground-level fuel will not accumulate to the point where flames will harm a stand.

Lacking specific guidance for the area and species in question, natural occurrence can suggest a frequency for controlled burns. In the fire-prone natural forests of western North America, natural fires in any one stand occur at approximately 10-20-year intervals (Houston, 1973; Kilgore and Taylor, 1979).

If the tree is not fire-resistant, a cleared area around the stem can prevent damage (partial weeding). Again, the process must be carefully evaluated and undertaken to insure a planned outcome. As a result of the danger involved, burns, of whatever type, should be undertaken by those with extensive experience.

Grazing

Grazing animals are a weed control measure (as discussed in Chapter 6) especially suitable for monocultures. Briefly, the variables are the types of animals employed, the stocking rate, duration and timing (seasonally or across the rotational life of a plantation). Best results are obtained where the animals are paired with the vegetation. As elaborated in Chapter 6, the tree species chosen also helps focus grazing animals on the task at hand.

As for the influence of animals on tree growth, the outcome is mixed. Some report no effect (Couto et al., 1994), others report compacted soils and slower tree growth (Bezkorowjanyj et al., 1993). Therefore, stocking rates are part of any animals-related decisions.

Pruning

Pruning is another costly undertaking. Due to this, the alternatives are of interest. When economically justified, hand pruning is viable. Where not, a good starting point (a nice DPC) are a tree species (or varieties) with pronounced canopies and clear stems. Such trees are known and include eucalyptus; from temperate South America, the less noticed and seldom planted *Araucaria araucana;* or the non-commercial America elm (see Photo 2.2, top).

Lacking species with inherently clear stems, light demanding species that shed branches quickly are often selected. This coupled with a high planting density, side shade, and an early thinning will give the desired results.

Where economics do not favor an early thinning, side branches must be tolerated. This strategy is associated with pine plantations in the southeastern USA. Here the trees are mainly valued as fiber (i.e., pulp or chips), sawmills receiving the small percentage of naturally occurring clear stems.

Thinning

A thinning regime is part of keeping an optimal growth rate. Some employ a wide spacing and tolerate weeds or prefer to incur weeding costs while avoiding the expense of a non-commercial thinning. The risks are with shorter, branchy stems unless the selected species has a clear trunk as a fundamental DPC. Realized, and not an inherent quality, tall, straight, clear wood requires dense planting. As monocultures offer fewer agroecological options, manual thinning becomes more urgent.

With high initial planting densities, some tools exist that make the most of an uneconomic situation. In-place killing can be cheaper than a non-commercial harvest, but may be limited to the early periods when the trees are smaller, growth rates high, and the living trees are less likely to be damaged or squashed by the toppling of those killed. With limited ecological alternatives, manual methods come more to the fore.

Row thinning

For commercial thinnings, the pattern of stem removal is a factor in maintaining stand quality. This can be on a row basis or through the selection of individual species. The row method is employed where (a) the costs of extraction are a concern, (b) there is a high degree of uniformity in quality and size with the stand, and, (c) due to crowding, harm will come to neighboring plants unless extraction avenues are in place. Photo 8.2 depicts a plantation ripe for a row thinning.

Photo 8.2 A crowded plantation requiring thinning to insure good growth rates. The photo was provided by the Massachusetts DEM.

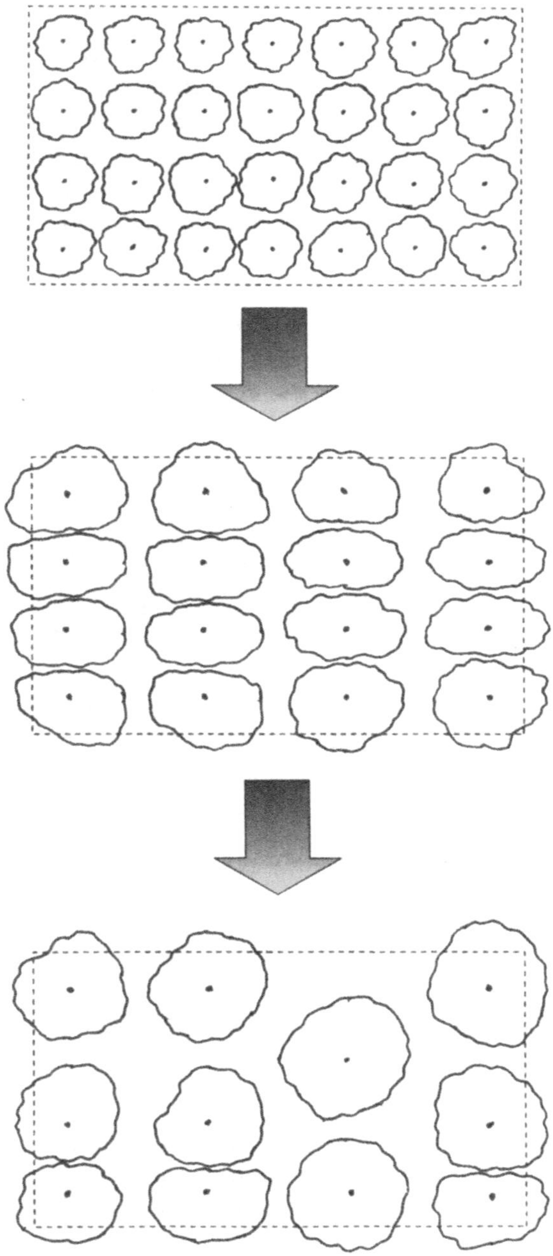

Fig. 8.1 The temporal cutting pattern for a monocultural stand where, in spaced-time periods, row extraction is followed by the removal of individual trees.

Individual plant thinning

In contrast, the individual method is favored where there are growth and/or quality differences between individual plants. The smaller, slower-growing trees are removed with compromises being made to insure that the overall spacing is more or less uniform. This method can be more costly and damaging in dense, early-stage plantations; made easier in more open, mature plantations that have been previously thinned.

Figure 8.1 shows a row-individual sequence where, for the first commercial thinning, the less costly, less selective row method was employed. For the second thinning, individual trees were removed utilizing the paths opened by the first extraction. Photo 8.3 is of an individually thinning-spruce plantation.

SUPPLEMENTARY ADDITIONS

In keeping with the looser definition of monoculture, supplementary additions can result in positive economic gains. Given the ecological simplicity, there are often many unoccupied niches that can be exploited and, with one species, it is easier to observe if an addition is having a negative impact.

Previously alluded to are rattan vines that extend into tree canopies and mushrooms and truffles that grow at or below ground level. Both can provide mid-rotation income without any wood growth penalty.

NOTABLE VARIATIONS

Under the realm of monoculture falls stands of uneven age. This offers considerable variation. The alternatives include:

(1) even aged, but with an uneven coppice;
(2) uneven, dual-aged (two-aged); and
(3) uneven, multi-aged.

These three categories offer a host of variations.

A discussion of uneven-aged stands was deferred to this latter section because (a) fully documented examples are missing from the literature, (b) there are a numerous variations off this theme, and, more importantly, (c) these variations also apply to species-rich systems. By putting the discussion of uneven-aged stands in

Photo 8.3 A spruce plantation recently thinned to an optimal density. Note the amount of canopy penetrating sunlight soon to be usurped by the standing trees.

monocultural form, the options are clearer. Throwing in multi-species complexity further complicates things. One should keep in mind that temporal alternatives also enhance the productive potential of biodiversity (Chapters 9 and 10).

In-plantation Coppices

In a version suggested by Pontey (1805), similar species are planted at the same time in a dense spacing and allowed to grow untouched for at least six years. When crowding begins to takes a toll on growth rates, alternating rows are cut. The resulting coppice is a second cutting cycle, suppressed until the first, non-coppice trees are removed. The sequence is shown in Figure 8.2.

Among the gains are better stem form for uncut trees, better in-house (ecosystem managed) weed control, and a better cost structure (assuming that thinning is less expensive than pruning). These gains may extend further where three or even four cutting cycles, all based on coppice regrowth, are incorporated into a rotation.

Fig. 8.2 The temporal series for a coppiced monoculture leading to an uneven stand. Although all are planted at the same time (top), cutting and subsequent coppice regrowth retards alternate rows. The gains include better weed control and a less branchy stem than from a wide spaced, uniform system.

Dual-aged Plantations

There are plantations designed with two age groupings. One option is based upon natural regeneration, a second, replanting with a few trees left from earlier rotations.

Shelter systems

Leaving a few trees from a first rotation, provides a few larger, more valuable logs, and, if needed, these serve as a seed source for the next generation. These are cut at the end of a second rotation.

An alternative sequence keeps trees from previous plantings (or cycles) until a multi-aged plantation exists. Figure 4.4 is one of many possible variations.

The large tree option

The example here originates from Graves (1911). Out of 100 trees initially planted

- 80 are taken in the first cut, 20 remaining and 80 (or fewer) replanted;
- the second cutting has all trees planted in the previous cycle being cut as are 8 from the first cycle, 12 large trees remain and 92 trees (or fewer) replanted;
- the third cycle has all from the previous cycle cut, 3 older remain and replanting again occurs;
- at the end, one or two large trees are harvested along with 98 or 99 smaller trees.

Figure 8.3 shows this sequence. This can be further summarized through the number of trees from the first planting (oldest age class) that remain after the

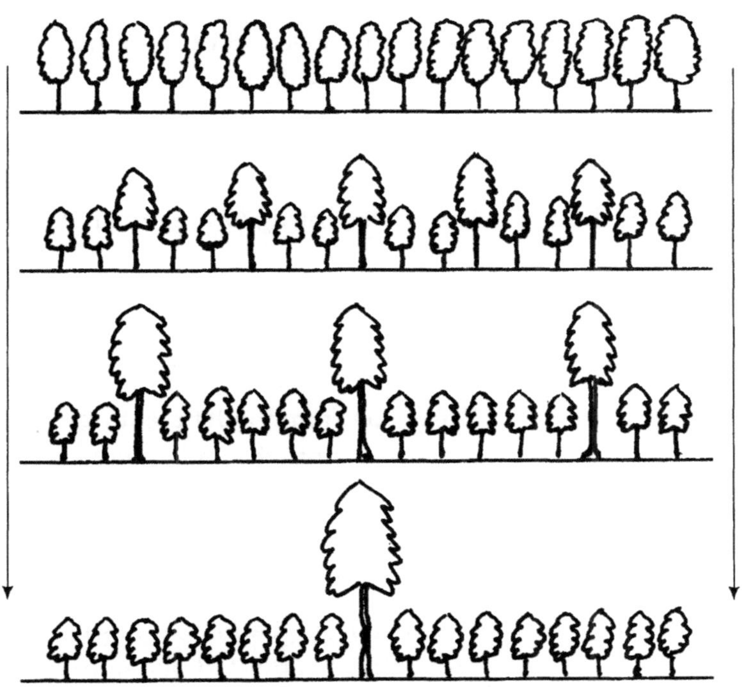

Fig. 8.3 An uneven-aged plantation sequence where a few trees from earlier rotation are permitted to grow to a large size. One of the ecological benefits is in the protection afforded newly planted trees by their older counterparts.

first cut	20
second cut	12
third cut	3
fourth cut	2
fifth cut	0

This system is designed to produce many pulp or pole stems and a few higher value saw or peeler logs. Other than planting, regeneration can be from coppice growth or through seeds from standing trees. The version, as presented in Figure 8.3, carried with it the assumption of a one-on-one planting ratio. Depending on species and the desired end result, this may not be optimal in all cases.

Multi-aged Plantations

Again, using the Graves (1911) example, another possible sequence is illustrated in Figure 8.4. The explanation has, out of 100 trees planted,

- 80 trees are removed at the first cut, 80 replanted and 20 remain;
- in the second cycle, 72 are removed (8 of these from the first planting), 72 or fewer are replanted;
- in the third, 69 are cut, (9 from the first cycle, 8 from the second, and 52 from the third), 69 are replanted;
- the fourth has 67 cut (1 from the first cycle, 9 from the second, 8 from the third and 49 from the third), 67 are replanted;
- at the end, all are cut.

This can be briefly summarized through the number of trees remaining in each age class (or after each cutting cycle) at each cut

first cut	20 (remain from the 1st planting)
second cut	12 (remain from the 1st planting)
	20 (remain from the 2nd planting)
third cut	3 (remain from the 1st planting)
	12 (remain from the 2nd planting)
	20 (remain from the 3rd planting)
fourth cut	2 (remain from the 1st planting)
	3 (remain from the 2nd planting)
	12 (remain from the 3rd planting)
	20 (remain from the 4th planting)
fifth cut	none remains and the cycle begins anew.

Fig. 8.4 A uneven-aged plantation that is more cumulative, i.e., a greater number of trees are carried across from the previous stage. The smallest trees in each of the four stages are from a new planting. This is also an example of a minimum interface design.

Given the space needs of the older plants, the number removed and planted need not be equal, but some lesser number may serve equally well.

There are some other prerequisites for success. One is that the value of older trees more than makes up for the increased management, the need for a selective harvest, and increased plant-on-plant competition. A minimum interface pattern (as shown in Figure 8.4) may have greater potential for improved stem form and might prove a key to success. As with all uneven-aged plantations, the gains could further improve if reformulated in multi-species form.

As a Successional Stage

Some have advocated employing a monocultural plantation as a prelude to or stepping stone for another ecosystem type. This is championed, in part, because most of the common plantation species are early successional.

If the plantation species is well chosen, site-soil improvements are possible (Lugo, 1997). With a suitable time lag, this sets the stage for a series plantation or the return of a natural forest. Allowing natural forest species to grow beneath an established plantation cover is one means to proceed quickly to a climax ecosystem.

Rather than waiting, an in-plantation forest (natural forest trees growing simultaneous with plantation species) can be more immediate although with fewer climax species. This concept is developed under notable variations of natural ecosystems (Chapter 12).

As a Form Shift

The possibility exists to implement an ordered plantation to serve as a starting point for a naturally regenerated, disarrayed, uneven-aged, single-species stand. Coppice regeneration is one way to do this, another is to sow seeds on a site that favors regeneration rather than succession.

Pine plantations are made more favorable, through grazing, for pine regeneration. The invading grasses and broadleaf trees are eaten by grazing animals while the ground litter disturbance caused by the animals encouraged seed-generated pine regrowth (Cooper, 1960).

LANDSCAPE CONSIDERATIONS

A discussion of inter-stand harvests, with broader implications that encompasses other silvicultural practices, is best deferred to Chapter 16. However, with landscapes of mixed monocultures, some topics, e.g., wood quality, unwanted fire, clean water, wildlife and social concerns, are of concern.

The edge effect is a wood quality issue. Increased horizontal light at the stand edge occasions undesired side branching and a proliferation of under canopy weeds. This can be regulated by having, as an immediate neighbor, a more nature, single-species stand or by employing a task-dedicated auxiliary systems (as with buffer or guide species).

Despite the comparatively limited ecological potential, monocultures have some positive DAPs. Thus, ecological chores can be accomplished, for better or worse, through monocultures. The water runoff from monocultural plantations or natural ecosystems is equally clean (Perry et al., 2001) while wildlife, if at home in a monoculture, may prefer one of native species and/or of mixed age.

In promoting wildlife and positive predator-prey relationships, much can be done through stand size and/or nearby auxiliary systems. Small sized stands with ample interface (both with monocultures of different tree species and with other, more biodiverse systems) are helpful in this regard.

Monoculture plantations in mixed landscapes can be positive additions or, if poorly placed, not economically or ecologically advantageous. In an already mentioned instance, exotic species can have a detrimental affect on nearby natural stands. In Uganda, Struhsaker et al. (1989) found that plantations of *Pinus caribrea* and *P. patula* caused severe mortality to close-at-hand *Newtonia buchananii* (with 100% mortality), *Lovoa swynnertonii* (90% mortality), and *Aningeria altissima* (45%). This is species specific, as the native tree *Celtis africana* experienced only a very slight (0.5%) reduction.

Although monocultures may have less environmental worth than other silvicultural options, the longer rotations in forestry and the lesser nutrient requirements of monocultural wood plantations (Fox, 2000) makes these a better alternative than their agricultural counterparts. This is especially true where short-rotation, high removal agriculture is the alternative.

CHAPTER 9

Bicultures

Biculture has greater potential, in terms of increased yields, revenue, ecological and environmental gains, than the monoculture. The increased potential is expressed through the plant-on-plant agrobionomic principles.

Increased yields can be systemwide or be focused around a single primary species. Improved wood quality, e.g., through taller, straighter stems, often translates into increased revenue. The ecological and environmental gains come through biodiversity which bring on risk reductions, expanded fauna habitat, etc. Despite the increase in intended biodiversity, these systems are generally species governed, not yet crossing the biodiversity line to ecosystem governance.

The advantages are such that a biculture should be considered before turning to a monoculture. A lack of study, confirmed tree pairings and complexity limits popularity. The underlying agrobionomic principles, if better understood from an application perspective, could encourage a greater utilization of bicultures.

REASONS FOR USE

Productive, revenue and quality gains are high on the list of reasons for in-field installation. Also favoring bicultures are the dynamics associated with well-directed biodiversity. These include in-stand weed control and the reduced risk from erosion, insects, diseases and winds. Ideally, good tree pairings can achieve any number of gains.

Favorable pairings are common in nature. Some found together in a single successional phase, others occurring as the result of a phase transition. In western North America, lodgepole pine can be found in association with grouse whorteberry or pine grass (Koch, 1996).

Across successional phases, oak can follow and overlap with pine. This occurs with both North American and European species (see Photo 4.1). Natural pairings are common and provide insight into which species are best intermixed for productive gain (see also Photos 2.1 and 9.2).

Productive Gains

A large body of theory, i.e., the agrobionomic principles, supports the productive gains from a biculture. This is coupled with convincing evidence from field trials.

For some pairings, weaker, per-plant growth is compensated for by a greater planting density and improved essential resource efficiency. Under such circumstances, the partial LERs, i.e. (Y_{ab}/Y_a) and (Y_{ba}/Y_b), will be less than one; the full LER total being greater than one (as in Figure 2.2).

Tree species may achieve better growth through a facilitative association. This is indicated when the partial LER for one of the contained species exceeds one; the full LER is also greater than one.

For a few species, a nurse tree may be a requirement for normal growth with a lot of potential candidate-species, e.g., Indian sandalwood requires a nearby second species (Rama Rao, 1911). There are other examples in India, a teak (1/3) and leucaena (2/3) intercrop resulted in teak growth greater than with a teak monoculture. In this case, the tree height was increased by 45% and the diameter increased over 70% in association with facilitative leucaena (Kumar et al.,1998).

In Hawaii, DeBell et al. (1985) noted that eucalyptus (*Eucalyptus saligna* and *E. grandis*) was 25% taller and 28% larger diameter when grown in close association with *Albizia falcataria*. The eucalyptus did not do as well when intercropped with *Acacia melanoxylon*, but still better than when grown alone.

Cases exists of mutual (cross) facilitation. For this, the partial LERs for both species are above one.

From Central America, both *Jacarana copaia* and *Vochysia guatemalensis* have higher growth rates in close association than apart (Petit and Montagnini, 2004). Additionally, Miller and Murray (1978) found that, at age 48, Douglas-fir stands averaged about 6% more in cubic volume when grown with red alder than as a monocrop. In addition, the harvest volume of the red alder was nearly equal to that of the Douglas-fir. This suggests an LER of around 2.06.

Positive results may require, in addition to a felicitous pairing, a species-suitable site. This is well established in theory. DeBell et al. (1987) found that *Eucalyptus saligna* with *Albizia falcataria* showed negative cross affects that seemed site-dependent. The sentiment being that, if well site-located, these and other combinations could exhibit the necessary above-one LER. It should be noted that, because of the minimum LER requirement (LER > 1.0), bicultures tend to be revenue oriented.

Quality Gains

A second species to improve stem form is entirely possible. Figure 9.1 shows how, in hemming in the primary species, tall straight trunks are realized. Ideally, the secondary, facilitative, stem-improving species will die from natural causes (e.g., shading) at the prescribed time. This skips the costly precommercial thinning associated with a stem-

Fig. 9.1 A second, shorter-lived, guide species employed to improve the stem form of the primary species. Ideally, internal dynamics (in the form of shade) will cause the form-improving guide tree to die off naturally.

improving, purely monocultural planting while, at the same time, releasing stored nutrients to the remaining trees.

Cost Reductions

Although most planned bicultures are revenue-oriented, there is the potential for cost-oriented designs. Savings result from (a) less need for manual weeding, (b) a threat reduction from insects and less need for insecticides, (c) the elimination of the precommercial thinning and/or, where applicable, (d) plant-on-plant nutrient facilitation and a reduction in fertilizer inputs. As always, the ideal couples increased revenue with cost reductions.

TYPES

In ecological terms, bicultures can be early stage succession with two or more pioneer species planted and harvested together. A temporal alternative will overlap an early successional species with a more shade resistant latter phase species. Staggered plantings and/or harvests is one means to exploit or accommodate the difference and gain productive advantage.

The bicultural types examined here are:-

(1) facilitative (a primary with a secondary species):
  (a) growth,
  (b) stem form,
(2) productive (two primary species):
  (a) simultaneous,
  (b) temporal overlapping.

Facilitative

A purely facilitative biculture has an unmarketable second species which directly contributes to the commercial success of the primary species and overall ecosystem. The secondary species seldom has bearing for a full cycle or rotation. If well selected, it will be suppressed or die naturally after it's facilitative task is accomplished.

Growth rate

Growth facilitation takes many forms. The pairings can be tree-tree, tree-scrub or tree-cover crop, whatever provides the best facilitative, in this case, growth affects.

A key mechanism, one that seems to transcend others, is the ability of a facilitative species to fix nitrogen. This assumes that the species being facilitated is nitrogen demanding. This far from exhausts the possibilities as species that accumulate and transfer other mineral resources could prove equally advantageous. Far less studied are those plants that furnish more than one mineral resource.

Weed control, the cycling of nutrients and soil protection are high on the list of ecological functions undertaken. All can contribute to increased growth in the primary species.

Stem form

Less well documented, but established in principle, is a second species to improve the stem form for the primary species. Dense spacing alone can produce straight, true, branch-free trunks. If a niche difference is not present, this can be costly in terms of lost overall growth, especially as an ecosystem matures and crowding takes its toll.

For tree-tree pairings, differences in growth rates and maximum height can be exploited to force a primary species to have tall, straight trunks. This can be accomplished through a facilitative, guide species

(a) that, although planted first, grows slow and is soon over-topped by a faster growing, primary species, or
(b) that is fast growing, but has limited height growth and will be over-topped by a shade tolerant, slow growing, potentially taller companion species.

In both cases, these associations can reduce side branching and decrease the need for manual thinning. If facilitative or with resource complementary, these associations can help in maintaining high growth rates for the primary species. Fruit trees, domesticated with a short stature, can be used. These have the added advantage of having well-reported DPCs.

Productive

With a productive species coupling, growth acceleration and improved stem form, are good side effects, however, the emphasis is on a high LER value and on harvesting two marketable species. The mechanisms are mostly with the efficient acquisition and partition of essential resources.

Ideally matched species will have resource complementarity across a range of sites. Although the existence of these site-independent favorable pairings is highly probable, few such favorable forestry two-somes has been confirmed.

Weaker forms of complementarity need to be site accommodated, e.g., may interrelate well only on resource rich locations, those with an abundance or a shortfall of one or two essential resources, or on soils with specific characteristics. This may be the more common situation. In contemporary forestry, little has been done to advance site-specific couplings. Much of the lore on what works comes from early sources. These are referenced throughout this chapter.

Simultaneous

Two resource compatible species can be planted and harvested together as a single rotation. The objective is to make the most of on-site resources and achieve a high LER. Although the plant-plant complementarity issue is paramount, there should be no, if any, increased cost with a simultaneous planting; no more than with a single species system. Multiple harvests, if needed, are less of a limitation and can be economically reconciled and profitably undertaken.

Temporal Overlaps

Temporal overlaps are variations of the uneven-aged monoculture, but in utilizing a second species. Some of the more prominent variations and starting points for temporal variations include

(1) where both species start together but, after the first commercial harvest (of one or some of both species), a few trees are left to be harvested in large diameter form;

(2) where a second species is planted midway in the first cycle and harvested after the first (as in a natural progression between successional phases);

(3) one species is left in place for two or more planting and harvest cycles of a second species.

The reasons for these lie in shortened rotations and higher LER values (through complementarity during the temporal overlap).

The variations of the uneven-aged plantation, as described in the previous chapter (as in Figure 8.4), might be better reformulated employing two or more species. This can confine species with age groups and/or management regimes (e.g., staggering the planting of facilitative and primary species) or mix species across the temporal plane (e.g., assembling within age groups both primary and secondary species). Alternatives add to the overall complexity but, given the right circumstances, temporal complexity might provide future ecological dividends.

MATCHING CHARACTERISTICS

A successful pairing is the process of matching desirable plant characteristics (DPCs). The idea being that the agroecological roles or needs are met by at least one of the component species. Amplifying upon the themes for guide species and co-productive species, some general guidelines have been proposed:

(1) a light demanding species can only be mixed with one that is shade tolerant if the light demander grows faster (Schlich, 1910; Yoshida and Kamitani, 1997);
(2) a slow growing light demander is only mixed with a faster shade tolerant species if
  (a) provided with help, e.g., pruning or thinning (Schenck, 1904), or
  (b) these are raised in groups, rather than individual spatial pattern (Schlich, 1910);
(3) when mixing shade resistant species, the growth rates should be equal or the slower one is protected from being dominated (Schlich, 1910); and
(4) two or more light demanding species should not be mixed, except
  (a) on very fertile, well-watered sites, or
  (b) when the light demanders are in short rotation (Schlich, 1910).

For some of these (i.e., 1, 2, and 3), short-lived guide and/or facilitative species are best. When these die, the essential resources contained are handed over to the primary species. These temporary additions also help with common chores, e.g., weed suppression and, as briefly covered in Chapter 4, help maintain optimal spacing; avoiding overplanting and the need for manual thinning.

These guidelines are far from thorough and are subject to numerous caveats and exceptions. Still, when considered with known inter-species mechanisms, they do pare the list of possibilities.

SPECIES POSSIBILITIES

The trees that are commonly found in agroforestry may be a starting point for biculture forestry. Poplar (northern Europe); *Nothofagus obliqua* (Southern Chile); *Faidherbia albida*, *Parkia biglobosa*, and *Vitellaria paradoxa* (parts of sub-Sahara Africa), and *Prosopis cineraria* (India) all flourish when in combination with crops (grow well and do not overly affect crop yields). If well paired, the DPCs should transfer to silviculture as either a primary or secondary species.

In past eras, oaks were recommended for various inter-tree mixes (Evelyn, 1664; Brown, 1861) and this continues to more recent times (Tseplyaev, 1965; Bainbridge, 1988). It should be noted that oaks are a large genus and not all may integrate well with other tree species.

The species mentioned earlier in this chapter also qualify as do those suggested under noteable variations in this chapter. Species growing in natural close proximity can, with understanding, be translated into plantation plantings. Photos 2.1, 4.2, and 9.2 show nature-suggested pairings. These far from exhaust the list of possibilities. Available data suggests an abundance of potential tree-on-tree couplings.

DPCs

The desirable characteristics of the two species system, to a large degree, revolves around complementarity. This can be a weak form, site and/or limiting resource specific, or in strong form, where the pairing does well despite the site or the limiting resource. Other characteristics are ranked as needed.

Primary DPCs

The DPCs of a primary species follow those of monoculture. These are

(1) site compatibility (including temperature, rainfall, soil, etc.)
(2) self-pruning,
(3) good stem form (depends on end use),
(4) good wood quality (again depends upon end use),
(5) ability to resist or repel insect pests and diseases,
(6) a tolerance to reasonable climatic variation, and
(7) pairs well with one or more species.

The major difference lies in the ability to overcome competition. In monoculture compatibility with other plants is not desired whereas,

Photo 9.1 A tree plantation with a second species, in this case, a facilitative covercrop.

in biculture, interplant competition is controlled by the two species in tandem.

A second major difference is fast initial or good overall growth. Important in modern monoculture, one of the species in a twosome (the primary or secondary species) should have this as a DPC.

The other differences are in roles assumed. Since a second species has assigned ecological duties, these will add to or augment those contributed by the primary species. For example, large, slow decaying leaves, with their triple role in weed reduction, erosion protection and moisture holding, can be a ground cover shed in prodigious amounts by a secondary species.

Secondary DPCs

All secondary species, whether a tree or some other plant type, have universal DPCs that apply in all bicultural situations. The more or less universal DPCs for secondary species are:

(a) niche compatible with the primary species;
(b) through above or below ground affects, crowds out or suppresses yield reducing weeds;
(c) leguminous;
(d) promotes positive insect dynamics; and
(e) for life below the primary species, shade resistance or, for trees taller than the main species, an open, light penetrating canopy.

These are added to those listed in Chapter 5.

In addition, there are specific requirements that apply in specific-use situations. DPCs that may apply to select sites are:

(f) drought resistance,
(g) hydraulic lift capacity,
(h) can find and make available certain mineral nutrients,
(i) can bio-remediate less than normal soils, or
(j) large leaves that shed frequently.

As specific-use DPCs, some fit a role and, as above, others a site need. The DPCs of two role playing plants, that of guide species and weed-suppressing cover crop, are enumerated below.

Guide species DPCs

As an internal ecosystem species, the following attributes are generally solicited:

(a) short-statured, e.g., of lesser maximum height than the primary species, short lived or can be commercially harvested when small (e.g., cinnamon or christmas trees);
(b) has dense canopy;
(c) collects more vertical light; and
(d) casts heavy, ground-level shade.

There are guide species that border ecosystems, plantation or otherwise and forestall the edge effect. In this, the DPCs are as above. However, there are purely system external DPCs that include

(a) collecting more horizontal light,
(b) having a high leaf area index (LAI),
(c) a thick contained root structure, and
(d) ability to counter the spread of organisms (insects diseases, etc.) that attack trees.

Guide trees need not be forest species, fruit and nut trees serve well in this role. Among the advantages, these are widely available, are often dwarf or of short stature, their DPCs are known, they have an immediate economic value and help sponsor wildlife. The output may be valued but, for sustainability sake, may be left unharvested. Green (1908) mentions fruit trees in plantation situations, possibly as external guide species; promoting good tree form and freeing a stand from the edge effect.

Ground cover DPCs

For perennial cover crops, the list of DPCs is:

(a) does not climb on taller plants,
(b) short stature,
(c) produces ground-level, heavy shade (through high biomass and/or large leaves),
(d) drought and frost resistant,
(e) fits within the temporal sequence of the primary crop,
(f) has allelopathic properties to prevent weed seed germination, and
(g) promotes a microclimate to speed the decay of residual vegetation.

In addition, the desirable, but optional, plant characteristics for covercrops are

(h) fixes nitrogen,
(i) is fire resistance,
(j) provides forage or an alternate harvest, and
(k) fights the most noxious weed species.

There are examples for many of the above. For a light demanding covercrop that is suppressed over time, Blanford (1925) found the southeast Asian herbaceous species, *Stephegyne diversifolia* or *Adina cordifolia*, served this purpose when planted with teak. In Africa, velvet bean provides an alternate harvest (beans), has the potential to cover the cost of establishment, and did well against speargrass; a dreaded, hard to control weed (Versteeg et al., 1998). Photo 9.1 shows unidentified trees with an anonymous cover species.

The above-listed attributes should also hold true when shrubs or trees are employed as cover crops. These plants are densely planted,

Photo 9.2 Trees growing in close harmony, in this case, yellow birch near hemlock. Although not definitive, naturally occurring pairings can point out species with bicultural potential.

usually through broadcast seeds, and rely upon complementarity with the primary species to rule these out as a competitive influence.

SPATIAL PATTERNS

To achieve the more involved goals, there is more reliance upon a spatial planting pattern. A well planned design will match species with site, helping to realize other ecological functions. The patterns can be individual (fine), based upon planting density, or can be coarse, relying on groupings to achieve purpose (as in Figure 3.4).

Primarily, spatial patterns are used to match the site with the species and to achieve ecological goals. Following the fundamental rule, the fine patterns bring to the fore the growth advantages of well-matched pairings.

The teak-leucaena pairing, mentioned earlier in the chapter, has planting density of 1/3 teak and 2/3 leucaena, maximizing the facilitative gains that the leucaena confers upon the teak. For this pairing, the obvious row arrangement is, in cross section, *...abbabbabba....*

Other than the fundamental rule, there is little to go on. For now, users must work from prior knowledge or what can be observed.

MANAGEMENT OPTIONS

As biocomplexity increases, some management options are lost, but far more arise. In a well-designed biculture, weed control should be enhanced by having two species. If critical, the second species, through a ranking of DPCs, may be specifically assigned this task.

As with monocultures, the other options are fire and grazing. As an added dimension, fire and grazing are also tools for managing a facilitative addition. Here again, DPCs in the primary and second species encourage these activities and their inherent gains.

Where there are two primary species, fire is seldom employed unless both species are fire-resistant. There are exceptions. Fire can be employed to hasten the demise of the second, non-resistant species once the facilitative tasks have been performed. This can happen in a stem-quality improving biculture where the primary species is fire-resistant and the secondary species is easily killed. It may be important that the second species not contribute to increased fire intensity.

Grazing is also a suppression tool where the secondary species is eliminated once it becomes a competitive threat. For this, the secondary species may be a shrub or a tree with eatable bark, easily eliminated by introducing hungry animals at a prescribed point in the rotational phase.

NOTEABLE VARIATIONS

Despite the lack of fully documented examples, bicultures are well represented in past writings. These, coupled with agroforestry observations, add to the body of thought on possible pairings.

More Suggested Pairings

Evelyn (1664) proposed both bicultures and multi-species plantings, elm with ash being one suggested twosome. Boutcher (1775) noted that, on coarse stoney soils, chestnut is best with ash and, on light sandy soils, larch with beech. With other soil types, single species systems were suggested.

Bryant (1871) has Norway spruce interplanted with poplar or willow. In this case, the poplar or willow follows the spruce by two years. Schlich (1910) has, again for planting in northern Europe, silver fir with spruce, silver fir with beech, beech with hornbeam, and the exotic Douglas-fir with beech. The strength in these includes the frost and drought protection that spruce provides to silver fir and the blowdown security that silver fir affords to the spruce.

Other pairings seem to favor a less equal relationships. Spruce with beech sanctions the spruce on wetter sites, the beech on drier areas. This combination, beech with spruce, if as touted, lends planting flexibility as well as insuring against growth declines during periods of high and low rainfall. Other couplings are designed around a single dominant species, e.g., where beech eclipses hornbeam on all sites.

Although western Europe drifted away from bio-complexity very early in the twentieth century, more recent cases arise in eastern Europe. Tseplyaev (1965) describes plantations of oak with spruce or hornbeam, oak with ash or alder, and oak with linden. These associations are described as being soil specific.

These only sample what had been an interesting line of development, one mostly lost to mainstream forestry. A lot occurred in eras when bicultures dominated thought and in-field examples were commonplace (Borthwick and MacDonald, 1913). From the breadth of what was once presented, one can only assume that soil-specific pairings are not hard to spot.

Agroforestry Examples

In the sphere of the biculture, those with the greatest wood producing potential are shade systems. Common examples are (as primary species) coffee and cocoa where, under the protection of overstory trees, the crops are protected and crop management costs reduced.

A shade canopy can easily consist of two or more high value timber species. As these plantations are in place for decades, time is not a factor and high valued hardwood can be an end goal with an early or continuous income stream provided by the crop understory.

If wood is secondary, these fall under the heading of agronomic agroforestry. If wood is primary, these fall under a forestry heading.

LANDSCAPE CONSIDERATIONS

Inter-stand placements are a landscape function and best discussed under a landscape heading (Chapter 16). Biculture, with improved biocomplexity and the potential for strong DAPs, can be ecologically self-governed and self-contained within the landscape. With this comes less reliance upon the landscape to provide the full measure of environmental protections.

Still, not all bicultures are completely framed for a strong, independent, environmental role and mutual cross-ecosystem protection may be a needed necessity, or a welcome addition. Well formulated nearby ecosystems can help reduce the range and magnitude of ecological threats when a biculture lacks the needed DAPs.

CHAPTER 10

Three-Plus Polycultures

Bicultures have more ecological potential than monocultures, it therefore follows that, with greater biodiversity, more ecological gains accrue. Three or more species (the three-plus polyculture), without getting into the sphere of the complex disarrayed ecosystem, is a large jump in biocomplexity.

In terms of governance, three-plus polycultures fall between species and ecosystem where, with some understanding of principles and practices, returns at multiple levels occur. This includes an enhanced ability to efficiently utilize essential resources and/or handle different types of risk.

In addition to more involved niche relationships, intricate spatial patterns come to the fore. This is not just with the basic ground-level patterns, but with vertical patterns and dimensions not found with the biculture.

REASONS FOR USE

High on the list of the gains from biodiversity are LER and log quality but, with more species and more ecological roles, the DAPs are solidified. As a consequence, this brings on a more risk-free system and in greater environmental harmony. In theory, both species and ecosystem governance can be exploited to their fullest. This may be feasible as their close cousin, the highly biocomplex agroforest, seems to live in the best of all worlds.

At the low end of this spectrum, the triculture, productive gains have been reported. A three species, mixed plantation from Central America, consisting of *Jacarana copaia, Vochysia guatemalensis,* and *Caloplyllum brasiliense,* had an LER estimated at 3.45. This is a well received value, made possible because the latter two species exhibited higher growth in association than as monocrops (Petit and Montagnini, 2004). The third, *Jacarana copaia*, may have more of a facilitative role.

Most three-plus plantations realize gains through greater biodiversity and through more complex plant-on-plant functions. In a case from Nigeria, a mix of forest tree-species (11+) were interplanted. Agricultural crops inaugurated the ecosystem, serving to accelerate it to a planned ending; as an enriched climax forest. Reportedly, the growth rates for two inclusive, high valued species, *Terminalia superba* and *Khaya ivorensis* were higher than if these same species were raised in a natural forest (Lowe and Ugbechie, 1975).

Early foresters recommended multi-species systems and provided a wealth of descriptions. Unfortunately, this approach fell from favor before scientific data was routinely presented. A few early samples are listed under noteable variations (this chapter).

If there is any one disadvantage, it lies in the lack of understanding of plant-plant relationships needed for full implementation. This may be less of a barrier than supposed and a shortage of examples may be due more to trends in forestry than a lack of overall understanding.

TYPES

With three-plus polycultures, the number of variations is exponential to the number of included species. In contrast to a species-equal plantation, differing harvest values, growth rates, and a lack of knowledge on competitive characteristics means that some species dominate, others have a lesser role.

As for the types of relationships, a single primary tree with multiple secondary species is the simplest form. A dual primary and multiple secondary intercrop is further along the biocomplexity road. Other variations are possible.

Spatial

Without guiding principles and a strong understanding of the individual species, it can be difficult to enter the realm of the three-plus polyculture. Access is often expedited, and plant-on-plant intricacies displaced, by expanding upon known bicultural principles, practices and pairings. Named, the guiding principles are:

(1) expanded primary,
(2) expanded secondary,

(3) merger,
(4) separate re-enforcing,
(5) transitional, and
(6) a primary species with a natural component.

These six approaches are illustrated in Figure 10.1 through 10.6. These should succeed if based on the rules for complementarity and around sound bicultural pairings. It also helps if

(1) there is little or no competition among multiple primary species and, if this occurs, competition is limited or eliminated through
    (a) plant-plant complementarity,
    (b) wide spacing, and/or
    (c) an interceding secondary (buffer) species;
(2) there is little or no competition between adjoining secondary and primary species; and

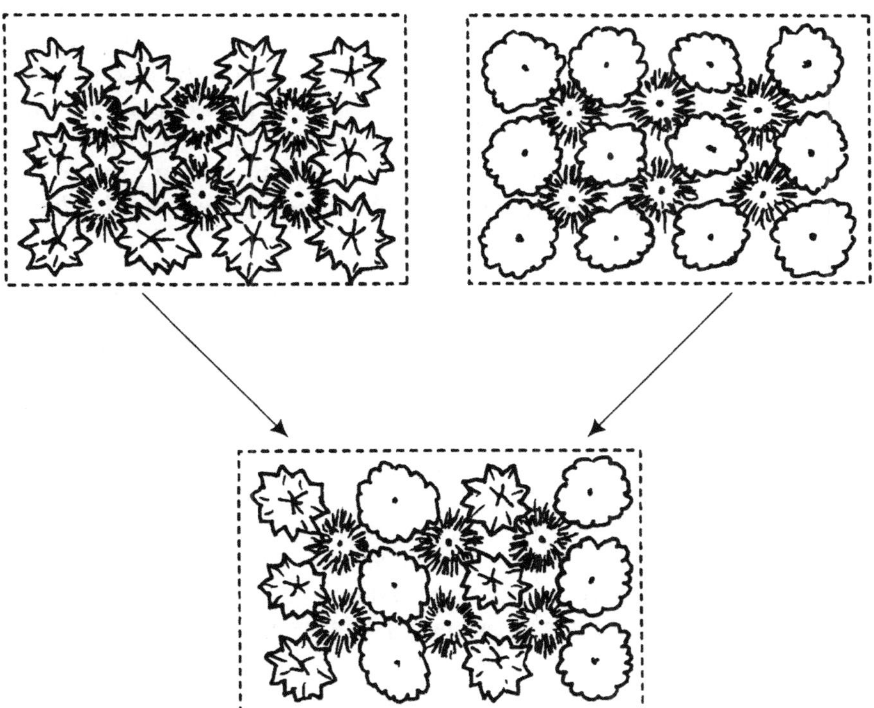

Fig. 10.1 An expanded primary approach where two ecosystems, each with a common secondary species, are merged. If the two primary species are competitive, wide spacing will ensures that inter-species competitiveness does not lessen growth.

(3) competition and suppression is permitted (or encouraged) between secondary species where these are part of the temporal planning.

Expanded primary

An ecologically successful biculture, with single primary species, can be expanded by adding a second primary species. This can be varietal, within a genus category, or employ an entirely different species. Figure 10.1 shows an inter-species pairing where two primary species, each coexisting well with a common secondary species, are combined.

Although nice, the two primary species need not possess plant-plant complementary, but both should be complementary with the facilitating (secondary) species. Without a requirement for inter-plant complementarity with primary species, there is greater flexibility for species selection and biodiversity.

If needed, the non-complementarity between the primary species can be managed through spacing. This is in line with established principles where complementarity is exploited through close spacing. Given the case where species *b* and species *d* are primary and competing, an interceding secondary species, species *c*, can fill a empty space and/or open niches (above and/or below ground). In cross-section, this is ...*bcdcb*... or, in more competitive cases, ...*bccdccb*....

Expanded secondary

A single primary species can be the focus of a three-plus polyculture, but more that one facilitative species may be added. Figure 10.2 shows where two secondary species, each coexisting well with a common primary species, are combined. The additions can be multiple guide species, multiple cover crops, a mix of a guide species and a cover crop, or facilitative plants with other roles and DPCs.

Merger

Merger takes two successful bicultures; merges or overlays these. Figure 10.3 merges two separate systems. As with all three-plus polycultures, this is only one of many patterns and arrangements that can be utilized.

The easiest to implement is where the two bicultures share a common primary or secondary species. Without common species, a spatial or temporal pattern can accommodate niche similarities (or competition) or space should be left between competing species.

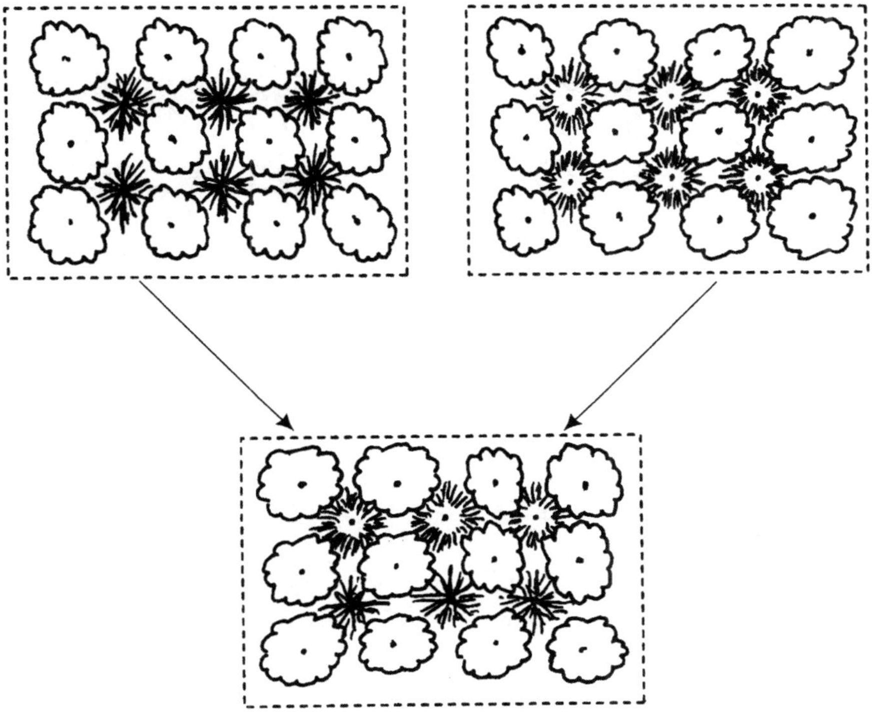

Fig. 10.2 An expanded secondary approach to three-plus polyculture. This combines two systems, both with a common primary species.

Through luck or understanding, there will be enough niche variability (complementarity) to make the merger effective. As with all species-governed polycultures, the exploitation of complementary is the guiding principle, i.e., species that coexist well together are placed near each other.

Separate re-enforcing

Instead of combining two bicultures, two systems can exist in close proximity, cross conferring many of the advantages of biodiversity while avoiding inter-system competition. The key is in having an intimate and inward association (with plenty of inter-system interface) and ample opportunity for ecological interaction. This can be with adjacent blocks, clumps or some other coarse pattern.

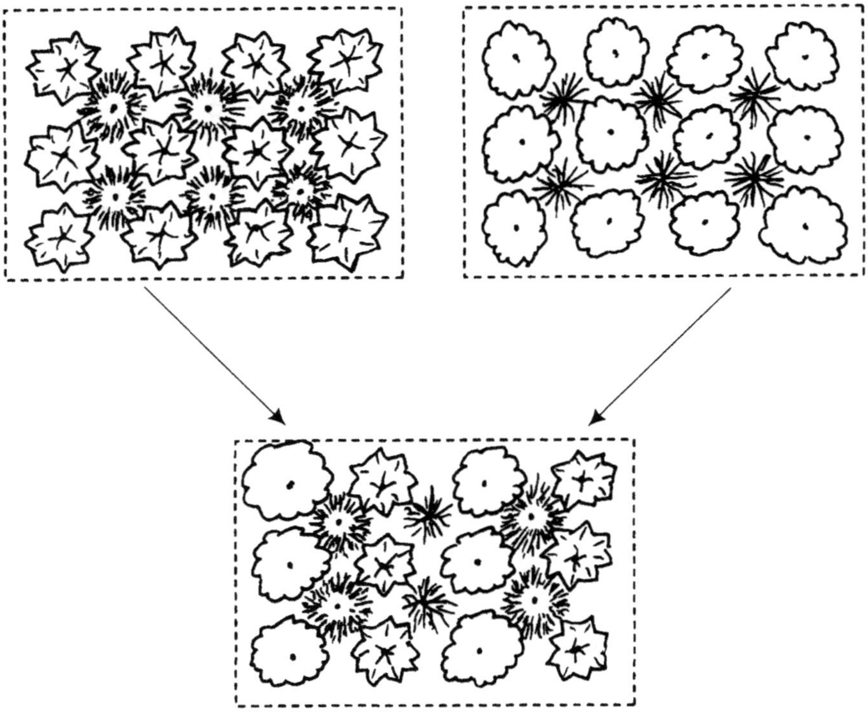

Fig. 10.3 The complete merger of two systems.

Figure 10.4 shows two systems abutting with plenty of interface. If there is no complementarity between the species in the unlike systems, an interceding guide species may be used to increase biodiversity, reduce interface competition, and bring the two systems closer together (physically and ecologically).

Transitional

Transition has the two ecosystems in close proximity along a common, high contact front. Figure 10.5 illustrates the intermingling of species where, upon association, these systems share a common, and mutually intrusive, interface. As systems grow in size, i.e., the amount of the interface increases, the depth of the intrusions should increase. As with other formulations, this is one of many spatial possibilities.

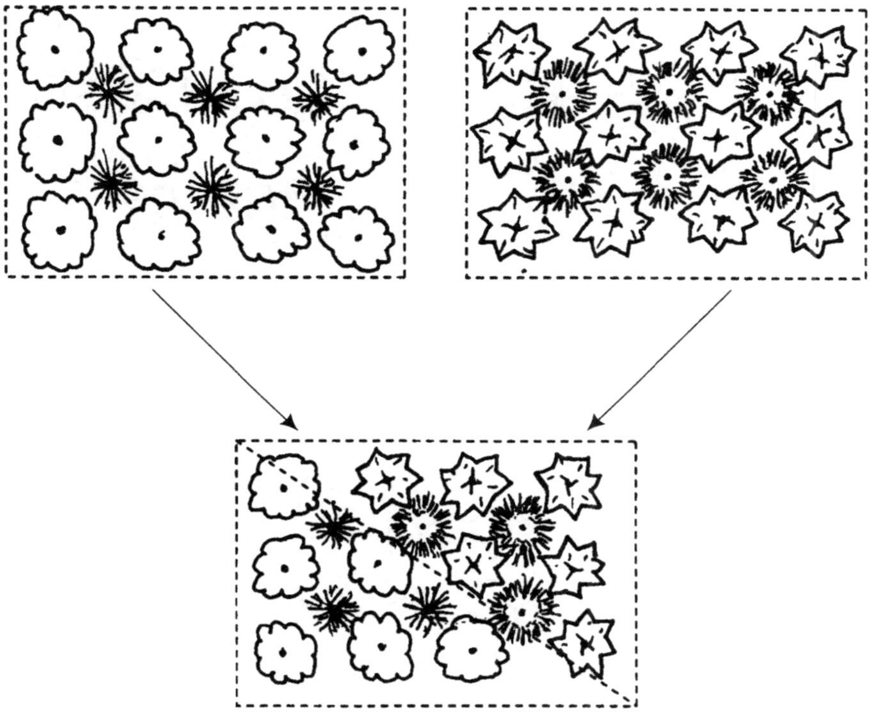

Fig. 10.4 Re-enforcing where two completely different systems are placed in close proximity. For the ecological gains to occur, the amount of inter-system interface should be high (as with the diagonal placements in this figure).

Natural component ecosystems

This form is based partially upon the principles of a complex agroecosystem with an ordered addition. For this, the parameters of density, diversity, and disarray are in place and are fully functioning and, as in Figure 10.6, an ordered system is superimposed.

There are a number of variants. Inclusive in this are:

(a) a biodiverse understory beneath an ordered upperstory plantation,

(b) an ordered understory below a disarrayed natural canopy, or

(c) strips or gaps of planted trees within the natural ecosystem.

The latter of these (c), in a strip version, is pictured in Photo 10.1 These options are discussed further under noteable variations in natural forest management (Chapter 12).

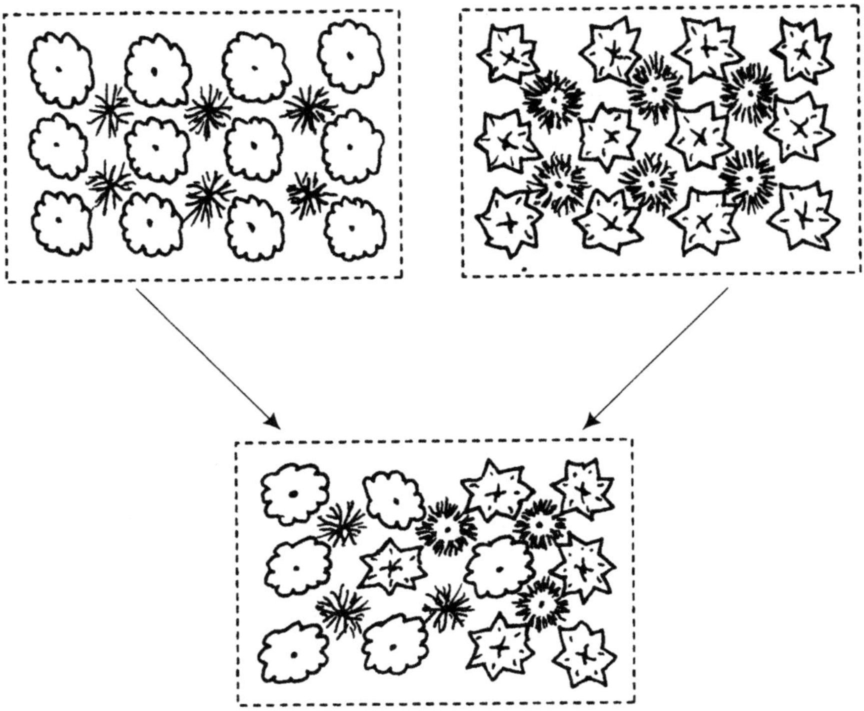

Fig. 10.5 A spatially transition approach to forming a three-plus polyculture, where a slight overlap occurs along the interface between the two systems.

Temporal

In addition to pure spatial designs, there are temporal alternatives, many expressed as variations of the uneven-aged stand (as proposed for monocultures, Chapter 8). Species complexity can add ecological depth and another form of biodiversity participation. Paired species can be kept within age groups (e.g., the simultaneous planting of primary species plus guide trees) or temporally staggered (e.g., where a facilitative species, planted first, sets the ecological stage for one or more primary species).

With definitive knowledge, simple options capture much of the ecological and economic potential. For example, relying on one guide species early in life of a plantation where a second continues in this role as the primary trees mature. Although the secondary species may have a temporal overlap, their roles in shaping the primary species (growth and stem form) do not overlap.

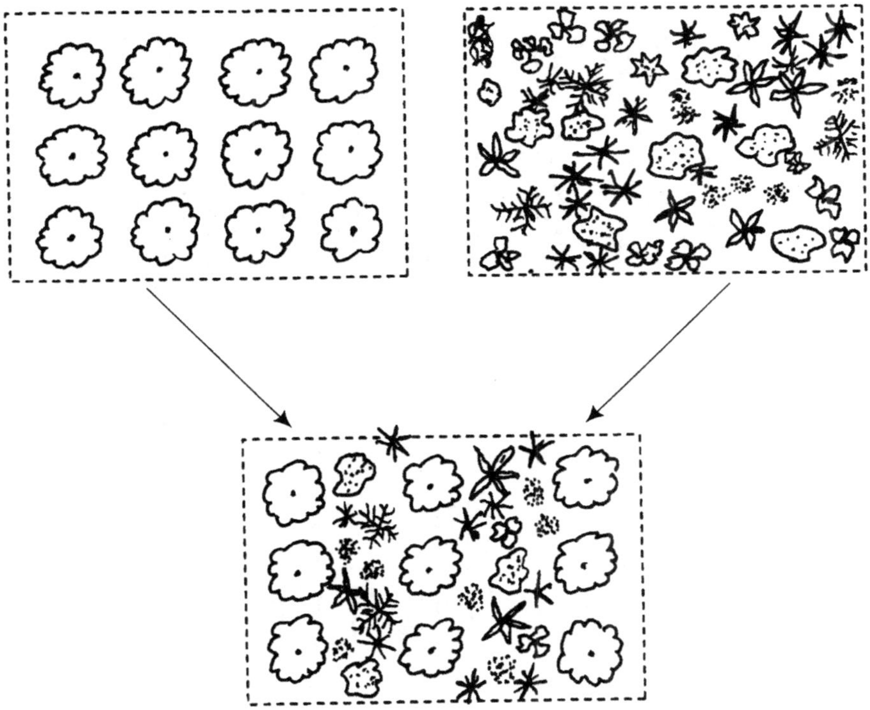

Fig. 10.6 A polyculture formed by merging a natural ecosystem with a monoculture. What results is an in-plantation forest.

Other possibilities include a role reversal between the primary and secondary species. Figure 10.7 shows one such temporal pattern where a shade tolerant, slower growing species (two or more) accompany a faster growing light demanding species. When the faster growing species is harvested, the understory species (one on more) flourish and these become the primary species. The sequence may continue where a heavily thinned stand is planted with light demanding trees.

DPCs

The DPCs of primary species in the three-plus polyculture remain much the same as with those for the monoculture. However, as the number of species increases, more ecological roles are filled by secondary species and a single ecosystem can accomplish more in terms of DAPs. This occurs because, as the number of cohabiting

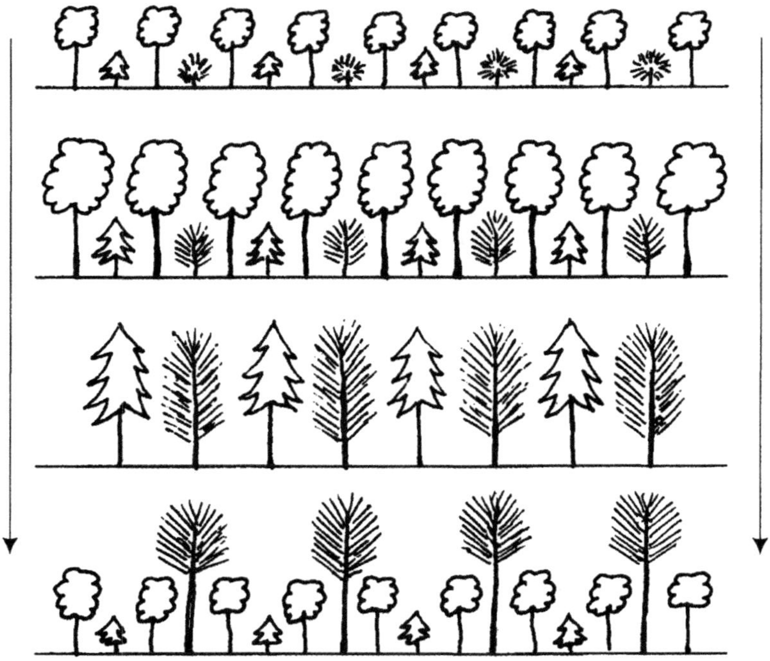

Fig. 10.7 A three-plus polyculture attained by employing a temporal (top to bottom) overlap.

species increases, so does the number of niches occupied. Fewer vacant niches means fewer opportunities for uninvited species (weeds) to establish, less unprotected soil to wash away, and smaller amounts of escaping unclaimed nutrients.

Without a fully-portrayed silvicultural study upon which to rest an explanation, one must turn to intercropping, specifically a maize-bean-squash mix. Foremost, one should not expect even growth. The harvest ratio being approximately 100% (maize), 50% (beans), and 10% (squash) yielding an LER of 1.6. The growth rates for cohabiting tree-species may well approximate this.

This intercrop also shows how the ecological roles of each species are partitioned while, at the same time, these overlap. The common assumption is that the nitrogen fixing capacity of the beans benefits the maize (although studies show a more intricate relationship befitting strong, site-independent complementarity). The biodiversity gains from the maize with bean portion includes better resource partitioning, e.g., marginal gains, along with weed and erosion protection.

Photo 10.1 A plantation with a natural ecosystem (an in-plantation forest).

This is not as simple as assigning a single task to each species. The squash, as a ground-hugging, large-leaf vine, is more for weed suppression than for the minor 10% output. However, the ecological functions of the bean and squash overlap as the beans augment the ecological input of the squash, adding erosion control, helping to retain in-soil moisture, and expanding the habitat potential for insect-eating insects.

Once understood, many tree-based three-plus polycultures may find similar multiple roles with shared inter-species responsibilities. The role differences between bicultures and tricultures may be subtle and difficult to statistically track where each small influence adds incrementally to the LER and/or risk reduction. For others, the gains for each component species can be large and unmisteakable.

As the number of plant species increases, ecosystems change governance. The ecological roles actively assigned to a species may be usurped by dynamics of the overall ecosystem. This is shown in natural forest management where, although component trees possess DPCs, these are, with the exception of those related to economic value, seldom enumerated. The same holds true as the amount of biocomplexity (along with density and disarray) increases in a planned and ordered stands.

PATTERNS

With more species, the possible patterns and arrangements are greatly expanded, so much so that even a partial discussion nets little. This is especially true given that there are no published studies that link pattern with outcome. The simplifying approaches (Figures 10.1 to 10.6) employ the standard bicultural patterns and can help initially with this linkage.

Growth complementarity is regulated by spacing and planting density. If one component threatens to overwhelm the primary species, fewer are planted. On a nitrogen poor site, the planting ratio may swing in favor of those plant species that fix nitrogen, while those that demand nitrogen are less prominent.

If resource and growth complementarity are in conflict (e.g., where a nitrogen-fixing secondary species is outgrowing a high-valued primary species), a different pattern can be tried (coarse rather than fine), the pattern arrangement altered and/or a temporal adjustment made.

A coarse pattern can include the application of self-suppression. Crowding can help to protect the soil and achieve other ecological tasks. An example is where a nitrogen-fixing secondary species (a) is densely planted. Once the ecological tasks are in place, self-suppression along with overstory shade would naturally thin this species. Species a can be a cover crop or short-lived tree.

MANAGEMENT OPTIONS

With a well-designed system, natural weeding, pruning and thinning can be easily accomplished through tree selection and accompanying ecological dynamics. As most of niches are appropriated, weeds should have little opportunity and be mostly non-interference.

Grazing animals are only added where an ecosystem is formulated for their inclusion. Fire is a blunt instrument and, given a high degree of biocomplexity and, unless very well planned, this can be more damaging than helpful. The only application may be in controlling weeds and recycling nutrients in even-aged stands of fire-resistant trees.

The other tasks, pruning and thinning, as with the biculture (see Figure 9.1), are a function of timing and of biocomplexity. If natural forest can accomplish these without direct intervention, so should a well-formulated three-plus polyculture. Therefore, biodiversity dynamics, rather than outside interventions, should a pre-planned driving force in management.

NOTEABLE VARIATIONS

The three-plus polyculture is mostly absent from recent literature. The examples cited here come from earlier periods and come without growth and yield data. The abundance of earlier-period examples demonstrates that multi-species systems were once widely accepted, but the techniques have been lost as later generations turned to simpler forms. Samples of those described are listed here by region.

Northeast USA

Roth (1902) suggested, on rocky, poor ground, planting a mix of oak, American chestnut, locust, elm, and maple. On better sites, Roth (1902) suggested white pine or some of the same species as suggested for planting on poor ground.

Prairie USA

Green (1908) lists some variations found, or suggested for use, in prairie regions of the central USA. The recommendations have:

- for porous moist soils (Iowa), white elm, black walnut, green ash, and hard maple accompanied by fruit trees;
- for porous moist soil (Minnesota), white willow, white elm, box elder, basswood, green ash with fruit trees;
- for dry prairie soils, green ash, box elder, white elm, and white willow accompanied by fruit trees; and
- for high prairie soils, cottonwood, white willow, box elder, white elm, and white ash, again accompanied by fruit trees.

The suggested fruit trees are wild plum, wild cherry, mulberry and/or juneberry. The fruit is not for direct harvest, but serves as decoy to keep birds from nearby crops. As suggested in earlier chapters, fruit trees do serve well as peripheral barriers, blocking the edge effect. Depending on the degree of integration, this system may also be classified as an agroforest (see Chapter 13).

It is also suggested that these plantations be encircled by poplar or by a mix of coniferous trees or tall shrubs (for inclusion as guide species). Again, this may be to block the edge effect. The poplar had been common in the southern prairies, in the northern regions, the recommended species are Norway pine, bull pine, red cedar and white spruce.

Europe

The ordered, three-plus polyculture seems to have a long history of use in England. Evelyn (1664) mentions oak thickened with beech, larch, Spanish chestnut, and some species of pine. Indications are that this is a well thoughtout grouping as Scotch fir (i.e., Scotch pine) is specifically not recommended for inclusion.

In early epochs, biocomplex systems were not out of the ordinary and a number are mentioned (Borthwick and MacDonald, 1913). A frequently mentioned forestry intercrop, one common in the early literature, is depicted below.

A descriptive example

Cobbett (1825) has mentioned that oak, beech, birch, elm, and ash "live harmoniously together" and have "different sorts of diet on the ground". In agroecological terms, this describes exploitable niche differences. These observations were carried into applications.

Brown (1861), Simpson (1903), and Nisbet (1905) are among those who describe a specific six-species intercrop. Of these, the Brown (1861) description is the most complete. This has a mix of hardwood and softwoods and has the hallmarks of an extended primary (Figure 10.1) and extended secondary (Figure 10.2) systems.

The overstory is of oak, ash, elm and sycamore. These are in rows with 6 m spacing. In one direction, the rows alternate oak and ash and elm and sycamore. In the other (perpendicular) direction, oak alternates with sycamore and ash alternates with elm.

The understory has the nurse (facilitative) species being larch and Scotch pine. These, at little over one meter (3 1/2 ft.) apart, have the Scotch pine further from a hardwood (1.8 m) with more larch than pine bordering the hardwood overstory.

The temporal sequence is flexible, with much depending upon growth rate. The general guidelines have the larch thinned at 10 years, the Scotch pine is removed at about 20-30 years. The ash is the first hardwood removed (at 30-40 years), followed by the sycamore (at 40 years), and the elm (at 45 years). The terminal, and therefore primary, species is oak.

Figure 10.8 illustrates parts of the sequence. The crowded early phase is shown in the upper cross-section. The intermediate phase, after the larch has been removed, shown in the center cross-section, the lower section has the hardwoods alone.

Economic outcome

From this system, a variety of woods are produced, starting with pulp quality larch, pole size pine, and saw logs of various hardwoods. The greater value (assuming that hardwoods have the greater worth) comes at the later stages of the rotation.

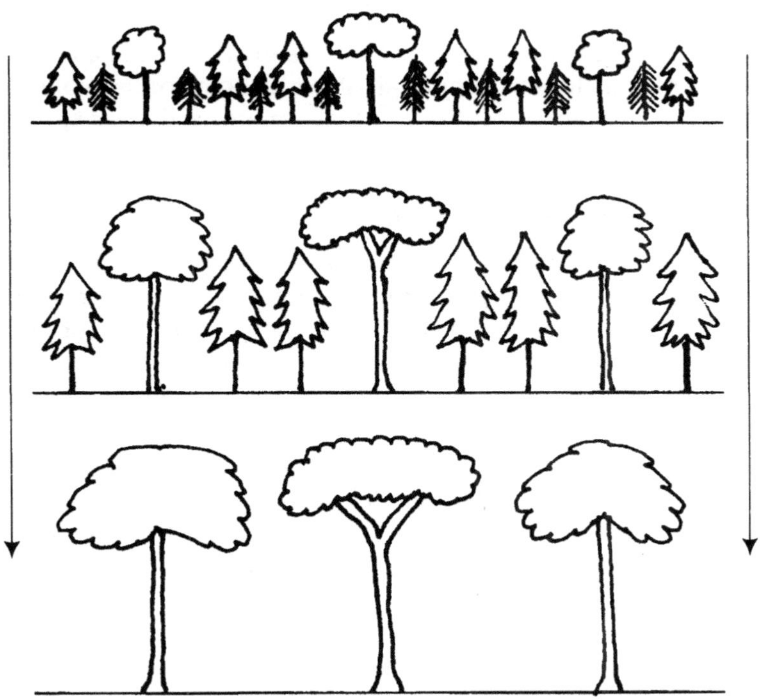

Fig. 10.8 The bio-complex example described in the text. The upper cross-section shows the crowded initial planting. The next (center) time period has the pine serving as guide for hardwood. The last period has hardwoods without the coniferous components. The growth rates are speculative as the available documentation does not go into detail.

As with many multi-species plantations, the continued harvests provide frequent and continued income stream. With good planning, this should equal or excel that attained from a series of monocultural plantings.

Agroecological dynamics

This system displays a lot of agroecology theory and serves to illustrate how biocomplexity is manipulated. The understory of larch and Scotch pine is clearly facilitative and confers a number of gains. Weed reduction is achieved while encouraging the hardwoods to produce tall straight stems. Since pines and larch do not stump sprout, this is not a management concern.

Also shown is a clever use of self-suppression. The close spacing (1.2 m) of the pine and larch favors early rapid growth and, once competition hits, the understory growth rates slow. The removal ordering of the larch and pine seems designed to reconcile varying growth rates and to encourage tall straight boles on all component species. These dynamics continue as the various hardwood species are harvested.

LANDSCAPE CONSIDERATIONS

With greater biodivesity and a more ecologically self-serving ecosystem, large areas containing a single system can exhibit considerable environmental friendliness. Better yet is to subdivide a large area, evenly mixing, in time and space, different rotation stages. The habitats gains in adding to the overall biocomplexity can indulge regional wildlife.

Even with intra-system biodiversity, cross-landscape associations can improve area ecological dynamics. As shown with the prairie USA examples, the inclusion of fruit trees can form positive associations, especially with nearby agroecosystems in a fragmented agricultural landscape.

CHAPTER 11

Taungyas

Taungyas fall under the heading of silvicultural agroforestry. What separates these from other silvicultural forms are the crops, treecrops or grasses (grazing) that are purposely placed under and in sequence with wood-producing trees. Although all silvicultural systems have temporal dimension, the taungya puts greater emphasis on what happens through time.

In the mid-1800's, the practice and the term taungya gained silvicultural prominence (Troup, 1940). The original application was in Southeast Asia where teak was planted with various crops. Despite expanded usage, the technique goes mostly unnoticed in the forestry literature. The possibilities go beyond the earliest and simpler applications to some highly intense, tree-crop combinations.

REASONS FOR USE

As a land-use alternative, taungyas have parallels in nature. The cropping phase, either planted before or simultaneous with tree plantings, is akin to the scrub phase that begins post-disturbance forest succession. Instead of fighting weeds, a simple taungya seeks to replace these with income-earning grasses (for animal forage) or agricultural crops.

The more complex taungyas mimic the succession of understory and overstory plants. This can be herbaceous (as with extended taungya) or with later successional elements (as with the multi-stage taungya). Here again, the overall ecological focus is on accommodation, rather than competition.

The earliest farmers clearly saw advantage in mixing trees and crops. In England, Evelyn (1664) mentions turnips for inaugurating the planting of hollies, laurels, yews and junipers. Also in England,

Browne (1832) has carrots, wurzel (a type of beet) and potatoes with tree plantings. Brown (1878) describes a French pine planting system that mixes pines and broomstraw. Schenck (1904) dates a German taungya of white oak with barley, oats, or summer rye as pre-1810. With countless crop and tree combinations, Asian versions have equally long histories (Menzies, 1988).

Whatever type, the taungya offers economic and ecological advantages. The income stream, that provided by crop harvests, helps early in the rotation when the trees are only an expense. Weeding and any fertilizers applied are paid for by the crops to the economic benefit of the trees. With greater inputs than have plantations sans crops, these systems are revenue oriented.

Gains also occur in being able to amortize the costs of the initial land preparation (e.g., clearing, plowing, etc.) between outputs (crops and wood). In addition, fire threat may be reduced, tree pruning serves to increase crop yields (through increased sunlight) as well as to improve stem form and, with a constant management presence, the trees are closely monitored (through IPM) for insect or disease outbreaks.

Among the disadvantages are: (a) a need for wider ranging expertise that encompasses agronomy and silviculture, (b) a possible reduction in per area crop yields, and (c) a reduction is the true cropping area due to the presence of the trees. Mechanized systems need smaller, more nimble, farm machinery and a machine accommodating tree spacing.

TYPES

Given the economic and ecological advantages, four taungya categories have evolved. These are:

(1) simple,

(2) extended,

(3) multi-stage, and

(4) end stage.

All contain, in some phase or form, an agricultural crop. This can be seasonal or perennial (as with an understory pasture).

As advocated for application in bicultures (see Chapter 9), fruit trees may be included. The classification difference lies in usage. In a non-taungya setting, fruiting species serve principally as guide trees, any fruit harvest being secondary. In the taungya, the fruit is actively sought and contributes to the economic success (e.g., RVT) for at least one of the successional phases.

Simple

The most basic is the simple taungya. For this, the taungya phase occurs only in the first year(s) after the trees are planted. In this equivalent of a successional scrub phase and the essential resources, those not exploited by the trees, are directed to crop production. After the first growing season or after canopy closure, the cropping phase is abandoned and some plantation type, either monoculture, biculture or a three-plus, ensues. Figure 11.1 shows the simple taungya.

This taungya comes to the fore where plantations are to be established and there farmers are willing to plant crops on the land for a short period. In land-short situations, there may be willingness on the part of farmers to clear the land and plant crops as well as trees.

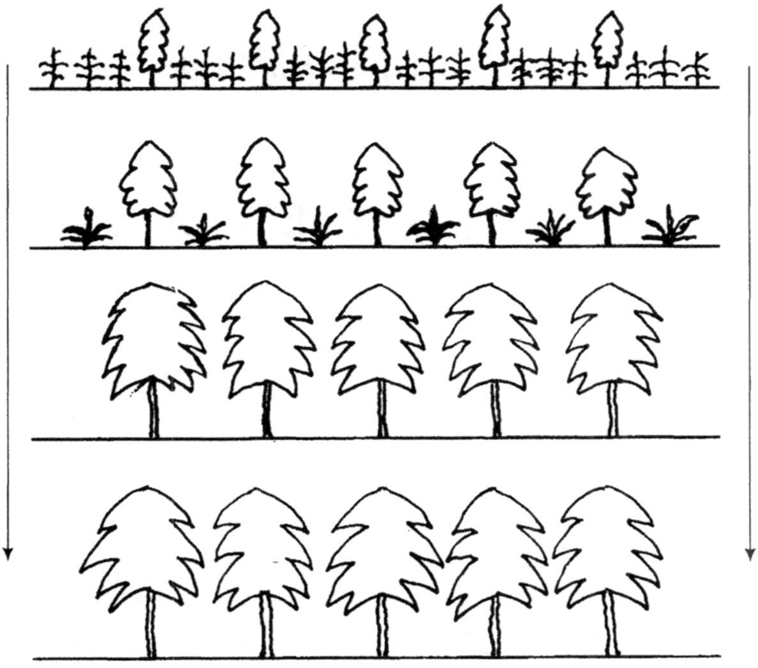

Fig. 11.1 A simple taungya where agricultural crops occurs only in the initial seasons, after which the system resorts to a tree-only plantation.

Where farmers are not willing to undertake this, a small stipend may persuade farmers to do the tree planting.

For these multi-participant taungyas, there are ample opportunities for mutual gains and for conflicts. For these to succeed, the roles and responsibilities of the participants (forests and farmers) must be clearly understood. In a case from Liberia, upland rice overgrew and set back co-planted teak when local farmers ignored a tree-crop spacing requirement.

Extended

The extended taungya has cropping or grazing occurring throughout an entire plantation sequence. As the competitive situation in the plantation changes, so should the crop type. In the latter stages, shade resistant perennials can be raised below the trees, commonly these involve shade resistant forage. Figure 11.2 shows cropping phases throughout a full rotation.

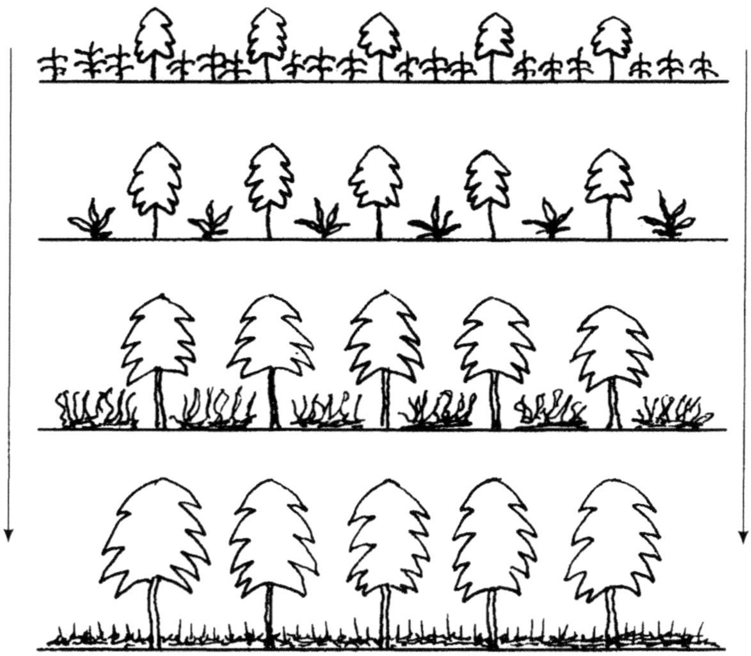

Fig. 11.2 An extended taungya where cropping or grazing occur throughout the entire rotation.

In contrast to the simple taungya, the extended taungya may serve better when agriculture and forestry are undertaken by a single management entity. The reason being the need for close coordination between the farm and forestry sectors.

Multi-stage

The multi-stage taungya form has a number of variations. The most common starts with a tree and taungya crop. After the crop phase, a second tree is planted and the first tree-species becomes the taungya species for the second planting. After the first tree is removed, a plantation of the second tree planting remains.

$$t_1\,c \dashrightarrow T_1\,t_2 \dashrightarrow T_2 \dashrightarrow \mathrm{I} \qquad (1)$$

This sequence is shown above where c is initial crop, t_1 the first tree planting. After some growth in species T_1, a second tree-species is planted (t_2) and, somewhere along the progression, T_1 is removed leaving the now taller T_2. T_1 and T_2 can be forestry species or the progression can have T_1 as a short duration treecrop and T_2 the forestry component. This version is shown in Figure 11.3.

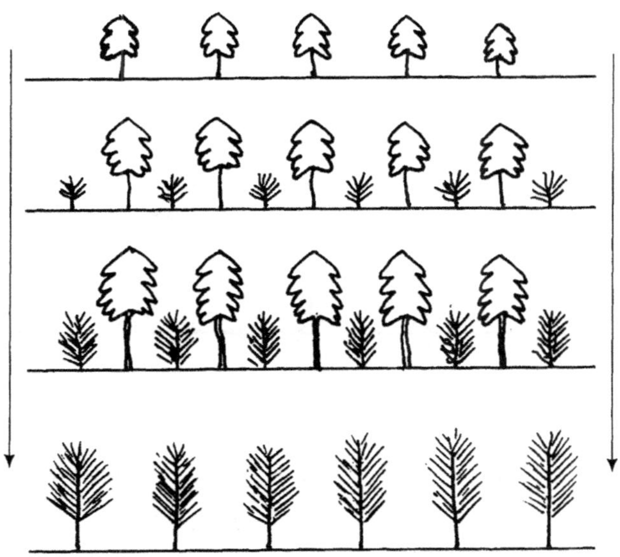

Fig. 11.3 A multi-stage taungya where the growth of one tree-species leads to another. This differs from a temporal overlap (see Figure 4.3) in that one species, usually the first woody plant, will be a treecrop or non-wood producing species.

Other variations may have overlapping sequences of three or more tree-species.

$$t_1\ c \dashrightarrow T_1\ t_2 \dashrightarrow T_2\ t_3 \dashrightarrow T_3 \dashrightarrow \mathrm{I} \qquad (2)$$

For this, T_1 and/or T_2 can be treecrops or a forestry species. These versions differ from an overlapping plantation sequence (Figure 4.3) in that these have one or more agricultural species. This can be a seasonal crop (c) and/or various treecrops. The ending plantation (T_3) is usually silvicultural.

End Stage

After the final thinning, when a few, widely spaced trees remain, the freed surplus resources can support a cropping or grazing phase. This is shown in Figure 11.4.

In a pine plantation, Clason and Robinson (2000) found that post-thinning grazing did not negatively affect tree growth and provided

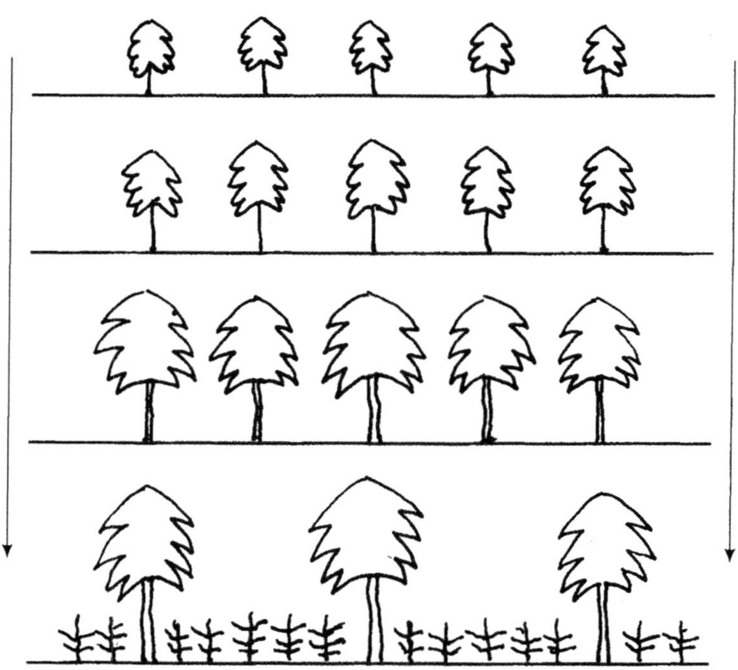

Fig. 11.4 An end-stage taungya where the space left after the final thinning is exploited for agricultural crops or grazing.

added income. Although not reported, biodiversity associated ecological benefits certainly do occur.

DPCs

The taungya, with the unique sets of ecological and the economic orientation, has DPC requirements that exclude many common monocultural species. In some aspects, there is more flexibility in understory species selection but, with regard to light, some severe limitations.

Primary Species

Early in the rotation, there are few limits on the tree-species as these are small and drawing few resources. Therefore, a simple taungya can include pines, eucalyptus and other competitive plantation species.

With more complex taungyas, the preference is with those species that tolerate or promote competition. Rather than selecting a primary species that can suppress or overcome weeds (as in the DPCs of monocultural species), trees are paired with crops such that positive niche dynamics and agrobionomic principles can be exploited. In temperate regions, poplars and paulownia are crop accommodating and are commonly employed in extended taungyas.

These, and other needs, brings a ranking of DPCs not found with other ordered forestry practices. These include:

(1) site compatibility,
(2) fast initial or overall growth (may not be important in the end phase of a multi-stage taungya),
(3) self pruning,
(4) good stem form (end-use dependent),
(5) good wood quality (again, end-use dependent),
(6) ability to resist damaging insects and diseases,
(7) a tolerance to reasonable climatic variation, and
(8) ability to accommodate competition (usually not important with simple taungyas) where,
    (a) there exists essential resource compatibility;
    (b) most roots are in sub-surface stratum; and/or
    (c) the tree has an open, light-penetrating or vertical, horizontal-light gathering canopy.

Secondary Species

As with primary species, the selection flexibility of the simple taungyas does not extend to other taungya forms. Shade resistance is not a well studied characteristic among crops, grasses and treecrops. For many common domestic varieties, years of breeding in light-rich environments may have suppressed the ability to survive as an understory plant (see discussion Chapter 5).

In some cases, crops can produce better in light constrained environment and, for some non-seasonal species, measures can be taken to improve the light-use efficiency in non-seasonal species. As an example of the latter, de Foresta et al. (1994) have advocated grafting a shade-resistant canopy onto the productive rootstock of cassava. Where understory shade resistance cannot be assured, a tree-species or variety should be chosen that has open, light penetrating canopy or where the canopy is compact and non-spreading.

PATTERNS

Given the nature of these systems, any coarse or mixed coarse-fine pattern is possible. Most common with mechanized agriculture are crop strips alternating with single lines of trees (a row pattern). Row spacing is set by machine width, allowing tractors and harvesters to traverse rows. The optimal tree density is initiated through intra-row distance.

For hand-planted, hand-harvested systems, more variation is encountered, including individual patterns. There is correspondingly more spacing flexibility where mechanization is absent.

Another consideration is row orientation. Normally, this is north-south, where the trees receive most of the horizontal (morning and afternoon light), the crops receive vertical (noon) light.

Where a crop thrives best with early morning light, a southeast-northwest (northern hemisphere) or northeast-southwest (southern hemisphere) orientation may be the best option. With this arrangement, a crop, energized by morning dew, can more effectively utilize direct sunlight. Later in the day, after drying, sun stress can adversely affect the crop and crop shading may be best.

Other orientations may satisfy specific needs with the caveat of reduced flexibility in rotating crops. Due to openness between young trees, orientation is less a concern in simple taungyas, more so with the extended versions.

MANAGEMENT

Taungyas benefit from traditional management inputs. Pruning is an ecological service provided by nature in well designed forestry plantations. In contrast, manual pruning may be the best option in many taungyas. As the agricultural activities provide funds and have available labor, this can prove a less economic or logistic burden as with pure forestry.

Other management methods can be equally unique to the goals of the taungya. IPM and some insect control options are examples of options that become economically viable because of the agricultural activity.

PLANTING

How trees are established has a bearing on the type of taungya chosen. Shortened rotations can be encouraged by planting trees in a larger form. Striplings and large stem plantings find favor in better quality cropland where mechanized transport and/or planting is available. Under these conditions, larger size stems may be cheaper to plant.

Large stem plantings can reduce the rotation by 20% and this has not gone unnoticed (see Photo 6.3). If an income stream is needed, the cropping period can be prolonged through spacing, pruning or selection of a shade-resistant crop. Extending the length of the taungya sequence is usually not the best course.

Pruning (Branch)

Manual pruning, may prove necessary and/or advantageous as, in opening the canopy, this (a) prolongs the cropping phase, (b) maintains higher understory yields, (c) allows farm machines to pass closer to the trees, while (d) improving the tree stem form. The costs are returned not only through wood quality, but through a lesser tree-crop interface, a proportionally greater cropping area and better understory productivity.

Pruning (Root)

Only with taungyas does root pruning come into play. It can be important to keep large surface roots away from crops. This is especially true with root crops such as potato or cassava.

Normal plowing often prunes roots and, lacking visual clues, this affect can go unnoticed. Where shallow plowing does not cut the offending root, this may have to be done through chisel plowing along the tree row. Figure 11.5 (center tree) has this affect.

It should be mentioned that cut and decaying tree roots can constitute a major vector for the transfer of nutrients from the trees to the crops. This reverses when the crops are harvested and crop residues decompose.

Root Collars

Large surface roots, an often unavoidable characteristic of some primary species, may be better dealt with through the installation of root collars. These are below ground barriers that force tree roots into deeper soil horizons. Wide plastic cylinders with a tree growing in the center is one, but not the only example.

These need not reach far, only deep enough to serve their intended purpose. Since tree roots commonly grow upward (with exceptions, e.g., the tree paulownia with a preponderance of deep roots) surface roots are not eliminated. Ideally, only the finer feeler roots will appear uppermost. These are more easily cut through plowing. Figure 11.5 (right-most tree) illustrates the effect of roots collars.

Fig. 11.5 Root management where (left tree) natural spread affects crops, (center tree) plowing redistributes roots to less effect, and (right tree) root collars change the spread profile and competitive relationship.

Grazing

Most taungyas, especially those in temperate and/or semi-arid regions, have grazing at some point in the rotation. Grasses, edible weeds or forage crops are encouraged or planted to feed domestic livestock. For best results, the usual advisories on stem bark protection, soil compacting and other negative consequences must be heeded. With the taungya, timing and stocking rates are adjusted for the best outcome.

Fire

Fire finds use in both cropping and silvopastoral systems. A fire-resistant primary species is normally employed, but an appropriate weeding method may permit light burns where the trees are susceptible. The frequency of fire depends on tree-species, climate, understory and other variables.

In addition to general reasons (see Chapter 6), fire removes old understory growth and promotes fresh new forage. In doing so, this better distributes animals, either by removing vegetative barriers to movement or, through the infuse of new forage growth, by erasing the memory of past gazing patterns (Shaver, 2002). As in any plantation, the dangers are weighed against the return and ability to safely undertake a burn.

NOTEABLE VARIATIONS

Among the myriad of possibilities is a combining of the simple and end-stage taungyas. This leads to other possibilities where crops are raised or the crops that end one taungya rotation are in place at the start of a new taungya series. As mentioned, grazing is so much as taungya component that silvopastorial systems and extended grazing are a large component of agroforestry.

LANDSCAPE CONSIDERATIONS

Given the revenue and cost advantages, taungyas find wide application. There is a direct link between land quality, the tree-species grown and the type of taungya.

On the more marginal sites, those better suited to grazing, simple taungyas with a fast-growing, weed-excluding species may be the better alternative. Where advanced cropping is part of taungya, quality farmland, a competition-accommodating tree-species, and large-stem mechanized planting are a highly advantageous combination.

In other ecological aspects, water, insects, wildlife, etc., the landscape will share many of the same risks as a landscape comprised forestry or agricultural monocultures. As with monoculture, layout

(See caption-next page)

Photo 11.1 Three taungya phases, (a) an early stage with crops, in this case, maize and poplar trees; (b) an extended type with grazing (forage) below trees; and (c) an end-stage taungya with sheep grazing under extensively thinned pines.

can mitigate much of this. Also with monocultures, widespread taungya use, depending upon types and layouts of the individual systems, could derive benefit from auxiliary technologies, i.e., neighboring systems that provide protection and ecological support. Insect-harboring natural strips, windbreaks, riparian buffers can come to the rescue when foretold dangers threaten. Chapter 16 discusses the layout options in greater detail.

CHAPTER 12

Natural Forest Management

The management of natural forests goes beyond the complex ordered ecosystem into the realm of the complex disarrayed ecosystem. Disarray, coupled with density and diversity, brings about system governance. The ecological forces unleashed supersede, but do not entirely replace, the plant-on-plant governance prominent in simpler ecosystems.

Invoking disarray departs from the fundamental patterns of ordered ecosystems, what survives disarray are the canopy orderings (midpoint or minimum interface) induced by management. Some coarse patterns (e.g., strip, clump and gap) also exist to direct ecosystems toward one or more specific objectives.

Another difference between complex ordered and complex disarrayed natural ecosystems lies in ecosystem membership. In natural ecosystems all the species present are coevolved with long-established ecological and successional dynamics.

With a natural ecosystem, specific silvicultural treatments can deviate from the forest norm. Enrichment is possible, e.g., natural species can be inserted that run against the successional flow, the populations of ecologically significant species can be increased or exotic species can be added.

TEMPORAL DYNAMICS

At times, silvicultural goals are best served in harnessing successional dynamics and keeping a forest on a successional path. This must be done so as not to be out of sync with silvicultural goals, environmental mandates and other non-timber aims.

Commonly, this means maintaining a climax forest, one defined as being species static where those species present exclude new additions

(Baker, 1934). This does not mean that the same plants remain in perpetually; only that those lost are succeeded by plants of the same species. Across time, with only minor differences in the species mix will actualize. In keeping the species mix static, a sustainable silvicultural treatment will speed the growth of the individual trees without changing its climax character.

Natural forest management is not always intra-successional, but can involve advancing from one phase to another. Nature may do this slowly, an appropriate silvicultural treatment can accelerate the processes.

Nature regresses through disturbance and practitioners can duplicate this, repositioning from an advanced phase to one that would normally have occurred earlier. If done with the severity of many natural traumas, then the result may be a scrub phase. If regression is undertaken in a controlled manner, then it is possible to make a controlled descend of one or more successional stages. The reason for any temporal movement (ahead or back) is to encourage high-value primary species (one or more) found in a particular phase.

TEMPORAL INTERVENTION

The temporal parameters of silvicultural activity are the rotation, cutting or management cycles and what may be termed the take. The latter may be circumscribed in terms of trees harvested (type, wood volume, age, and/or size) or in terms of those remaining (again; type, volume, age, and/or size).

These concepts, rotation, cycle and cut, are illustrated with a prescription proposed for the climax pine forests in western North America (Langston, 1995). Although not followed, it serves well as an example.

The proposed treatment has a rotation of 180 years with three cutting cycles. The first cycle takes 80% of the standing volume, leaving 20%. The second, at 60 years, is based on growth having doubled the standing volume, where 80% is again cut. The third cycle, at 120 years, cuts all the mature trees, leaving only the regeneration that began growing in the first cycle. The rotation starts again with the first cutting at 180 years. This is illustrated in Figure 12.1.

A silvicultural treatment (either rotation of cycle) is based upon three temporal occurrences. These are:

(1) the starting ecosystem,

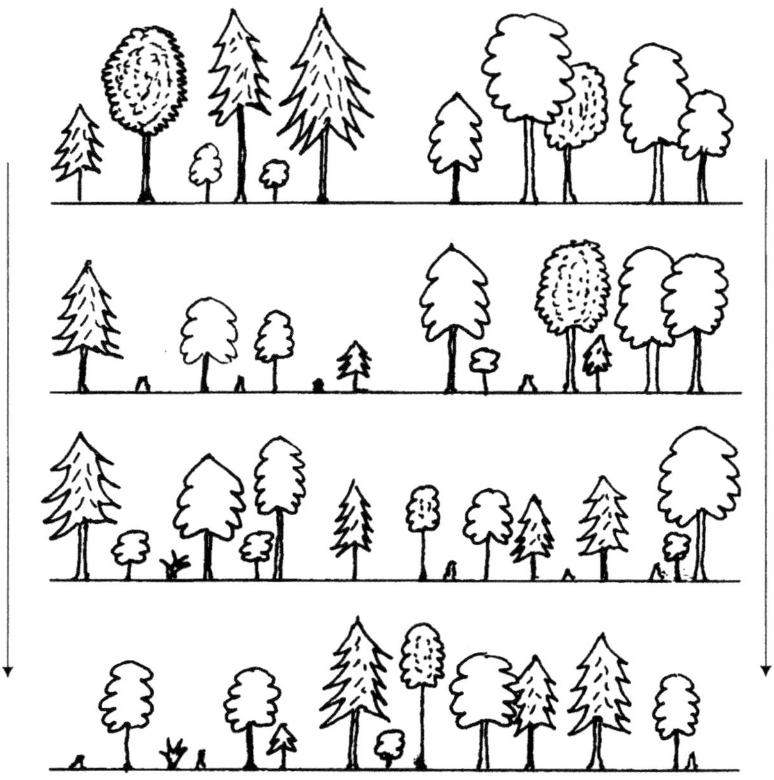

Fig. 12.1 Shown are the cutting cycles and rotation in a natural ecosystem. From the top-down, there are three cutting cycles. A rotation is concluded when all the harvestable trees that were standing initially have been cut.

(2) what is desired as an ending and/or lasting ecosystem, and

(3) what happens in transition.

A starting point can be a natural, climatic forest, as can be ending or lasting phase. What lies between a selective harvest, can take a few trees, or the full sequence of successional phases, each harvested in turn. Others starting, endings and transitions are possible.

What complicates silviculture is that the cycles can be more or less uniform, each contributing to a lasting forest state, or each cycle, because of changing socioeconomics, can vary in objective and outcome. One rationale for a mid-term change is that the high valued species and/or larger trees have been cut and management must focus on that which remains.

The Starting Ecosystem

There are a number of starting phases in natural forest management. As presented, these are: (a) scrub, (b) early phase successional, (c) a latter stage successional, or (d) the climax phase. These phases may be ecologically intact and pure across a stand or have experienced disturbance that mixes phases within one forest tract.

The Ending or Lasting Ecosystem

Since forests do not have clear starting or ending points, the ending point can be any one of the four successional phases. Most often, the ending phase may define the forest condition that is desired into the foreseeable future. This can be further refined where the sought outcome is more even-aged than nature intended. The same can occur in naturally disturbed stands without human intervention.

Another outcome is a more species uniform stand. The hope is that the resulting forest ecosystem will be of higher commercial (stumpage) value.

Transitions

How one gets from the starting phase to the lasting ecosystem is the essence of natural silviculture. Some of these transitions are well studied and name designated, others less recognized. The passage from a less valuable mix of trees to one of greater commercial interest may involve a shift from (a) a climax forest to a latter succession, (b) a mid or latter succession to a climax, or even (c) a climax to an early phase forest.

STANDS

A stand is a sub-ecosystem clearly distinguished by its structure (Simon, 2001). If linked by an intended amount of interaction with neighboring stand, these may constitute a management unit.

A large, homogeneous, uniform forest, one without distinguishing sub-characteristics, qualifies as a stand. This is not the case when a forest consists of a mix of successional stages where each successional stage, and the resulting gap or patch mosaic, is part of one large, interrelated, non-uniform ecosystem. Such a forest, subdivided in this way, can have many stands, each a management unit, or, if strong inter-stand ecological interactions exist, one or a few management units that bridge many stands.

Photo 12.1 A small clearcut in a climax forest.

The customary ecological interaction, that which links segments and makes these a single management unit, is seed dispersal. Other agroecological interactions (e.g., wind shelter) or environmental concerns (e.g., wildlife) might also be included. The degree of homogeneity (uniformity) or heterogeneity (non-uniformity) constitutes stand texture.

Uniform Stands

For silviculture in many regions, stands are more or less homogeneous, cutting individual trees in a individual fine pattern (as seen in Figure 3.4). These are managed utilizing less intrusive silvicultural treatments, e.g., the removal of individual trees, to maintain the successional dynamics within a closed-canopy ecosystem.

Non-uniform Stands

The option exist to have patch-oriented management where openings, the result of small or large clearcuts, are large enough to change the successional order. Instead of closed-canopied, climax forest, the patches are of different successional stages.

The coarse (Figure 3.4), rather than the fine, patterns come into play. Any coarse pattern, and subsequent arrangement, is possible. Gaps or strips are commonly found. A patch or gap ecosystem may be the ecological norm by mimicking the historical record of a forest (discussed in depth in Chapter 14).

To achieve the desired result, the size of an opening determines the type of regeneration. The presence of nearby mother trees (to seed the newly created openings), along with other localized influences, also create a successional changeover.

Opening size, along with height and density of surrounding stands, determines the amount of sunlight and this favors shade-tolerant (later successional) or light-demanding (early successional) species. With small gaps come shade-tolerant trees, large openings encourage light-demanders (Schnitzer and Carson, 2001; Kneeshaw and Bergeron, 1998). The second requirement, already mentioned, is for nearby standing seed trees of the desired species (Greene and Johnson, 1996; Tuomela et al., 1996).

Other factors can predetermine the regeneration. Undisturbed leaf litter is more conducive to shade-tolerant species, whereas ground disturbance favors earlier successional, light-demanders (Molofsky and Angspurger, 1992; Putz, 1983). If these variables are not closely controlled, an unwanted scrub stage may result (Fredericksen and Pariona, 2002).

Fire can guide the regeneration in fire-prone ecosystems. Tree falls and dead woody material in natural gaps is conducive to a fire hotspot. This kills weeds, favoring the regeneration of fire-specific trees. In the western USA, this brings forth serotinous pines (White, 1985) or redwoods (Stephens et al., 1999).

One must not forget stump sprouts. In a tropical forest, Hartshorn (1989) found regeneration from stumps in a cleared area was substantial (40-70% of the trees cut regenerated). This accounted for about 13% of the total regeneration. This can be employed as a regeneration tool or interfere with a planned successional regression.

Using preplanned disturbance regimes, some treatment strategies have formulated. Two discussed here are the Swiss and Bavarian gap systems.

Swiss

Over larger stands, there is the option of having regularly distributed clearcuts, forming gaps of less than 0.5 ha. This follows the standard prescription of small gap size and adjacent seed trees to produce climax regrowth. This may be best where the topography of an area is constant, site differences (e.g., soil structure, nutrient base, etc.) small, and the desired regeneration can be assured. Topographically, these can be found on mountainsides or on flood plains.

Bavarian

The Bavarian system favors a few select species. In the original form, pure patches of spruce (*Picea abies*) are encouraged, other species are in mixed groups (Simon, 2001).

In other regions where the preconditions can be met, this approach has sub-stands of the more value species interspersed with those containing greater biodiversity, but less-valuable species. To fill the regeneration requirement, seed trees may be left to supply the regeneration along with suitable gap size and disturbance regimen to encourage the desired species.

Following general land-use guidelines (as in Chapter 6), it is better to locate a species where it grows best. As such, the Bavarian system can be topographically dependent. In inter-stand form, the Bavarian system has landscape-wide ramifications (discussed under landscape, Chapter 16).

TREATMENTS (PRESCRIPTIONS)

Silvicultural treatments are a means to direct natural ecosystems along a chosen path. In agroecological terms, treatments or prescriptions harness the dynamics of the natural ecosystem for productive and management purpose. Expanding on that previously presented, silvicultural treatments can contain the following elements:

(a) a starting mechanism,
(b) silvicultural goals (tree sizes, ages, growth rates, etc.),
(c) a means to achieve this, and
(d) other considerations (e.g., conservation of wildlife, clean water, etc.).

A silvicultural treatment can:

(a) convert an existing ecosystem to the one desired (successional and/or one of age and/or species uniformity);
(b) change from one set of objectives to another (e.g., species composition and/or log size); and/or
(c) keep on a previously established course.

These can be employed to maintain a single overall phase (as with uniform stands) or a number of phases (as with non-uniform stands). To advance various goals (wood and other needs), a number of silvicultural treatments or prescriptions have unfolded. These can be roughly ordered as to their intrusiveness. Starting with those least likely to produce dramatic change, these are:

(1) sparse and infrequent harvests;
(2) damage-salvage cuttings;
(3) species-oriented sequences;
(4) senility cuttings;
(5) selective-shelterwood thinnings;
  (a) light crown (the French method),
  (b) heavy crown (the Danish method),
  (c) from below (the German method),
(6) liberation cuttings;
  (a) advancing,
  (b) reverting.

Sparse and Infrequent Removals

The cut in a previous untouched or a forest can be infrequent with very little standing volume taken. In the Amazon region, Southgate (1998) reported a cut of about two trees per ha. In the previously untouched tropical rainforests of Liberia, the initial take was one tree for each two to eight ha.

The harvest of only a few, large, high-value trees maintains an economically useful ecosystem without seriously altering the present or future character of the forest. The overall system is little altered because, the intermediate and small diameter trees remain to restore what had been. Continued sparse cuttings at lengthy intervals (usually decades) is an ecologically viable option with undisputed sustainability.

Damage-Salvage Cuttings

The destruction that naturally occurs can be the basis for wood removal. Whenever trees are naturally damaged, such that survival is no longer guaranteed, these trees can be harvested. This might be with large intense burns where charred trunks remain or with extensive areas of blowdown.

The assumption is that the now dead logs are wasted and that a large percentage of the wood can be taken with minimal ecological impact. This is not always true, the taking of deadwood from small gaps can interfere with the ecology of an ecosystem. Fallen trees are an environmental plus for a variety of wild animals and do contribute to the overall successional character of the ecosystem (see Chapter 14). For larger areas, the assumption may be true, but there may be environmental costs and, unless care is taken on logging, damage to the land will occur (Gravitz, 2003).

Besides possible environmental sacrifice, there is the unpredictability of meeting industry demands through disturbance. This is especially true when risk reduction is implemented as part of a treatment (as discussed in Chapter 7).

Salvage as policy has a major disadvantage in that commercial interests receive monetary gains from a burned forest. This is a powerful incentive for major conflagrations.

Species-oriented Sequences

One European system that has not attracted much notice is an irregular shelterwood or, more aptly termed, a species directed harvest (Troup, 1928). The main feature is that a different species is taken in each cutting cycle. The cycles are years, if not decades, apart.

In tropical forests, with greater biodiversity, one, two or three species may be targeted in each harvest. As with sparse and infrequent harvests, only the larger diameter trees are taken. The remaining trees should keep the species mix more or less constant. Besides possible marketing gains, this is viewed as a strategy to protect tropical rainforests from overexploitation (Rice et al., 1997a).

Senility Harvests

A form of harvest has been advocated by Seydack (1995) and Seydack et al. (1995) where the trees are cut prior to natural morality. The criterion for cutting is crown dieback. For each species, when a certain percent of dieback occurs, so does harvest.

The advantages of this approach are in natural compatibility and management simplicity. The disadvantages are that more in-tree defects and a set species mix must be accepted, there being little opportunity for an overall improvement in wood quality and in the species mix (as through culling). Also, as growth rates usually decline with age, this treatment may not give optimal yields.

Yields may be improved when, instead of crown dieback as the determining factor, growth rates are monitored. When these fall below a set level, harvest occurs. This would support a more dynamic and more intrusive stand than the premortality version. Due to data needs, this growth rate comparison method is more difficult to implement.

Either form, that based on crown dieback or declining growth rates, carries the assumption that a climax forest with climax species is economically more valuable. More intrusive methods are needed to bring to the fore those species encountered in earlier successional stages.

Selective Harvest-Shelterwood

An inverse relationship exists between shelterwood and selective harvest treatments. Whereas selective harvests look at the number, size and type of trees cut, a shelterwood looks at the number and species regenerated. This adds a sustainability factor to the harvest method. In this case, sustainability is not in continued high yearly growth rates, but tree regeneration (and high growth rates) of the species desired.

As discussed, the more intrusive treatments (those producing patch and gap, non-uniform stands) can favor earlier successional, light-demanding, species. A continuation of natural dynamics, one favoring shade-tolerant, latter successional trees, can be assured by harvesting scattered individuals, rather than groups of species.

Within this framework, a lot can be done. A well chosen selective cut can be employed to change the successional phase, the species mix, quality (through culling), and/or even the age distribution of the trees. In addition to overall treatment, the selective harvest-shelterwood dichotomy (and which successional phase is countenanced) is evoked through the intensity of each treatment (i.e., via a light or heavy cut).

With this in mind, different selective thinning methods have been proposed. Figure 12.2 shows the treatments, starting with an untouched forest as basis for comparison.

Light crown thinning (French)

This method removes upper level, lower value trees in mixed stands thereby hastening the development of more valuable species. The goal

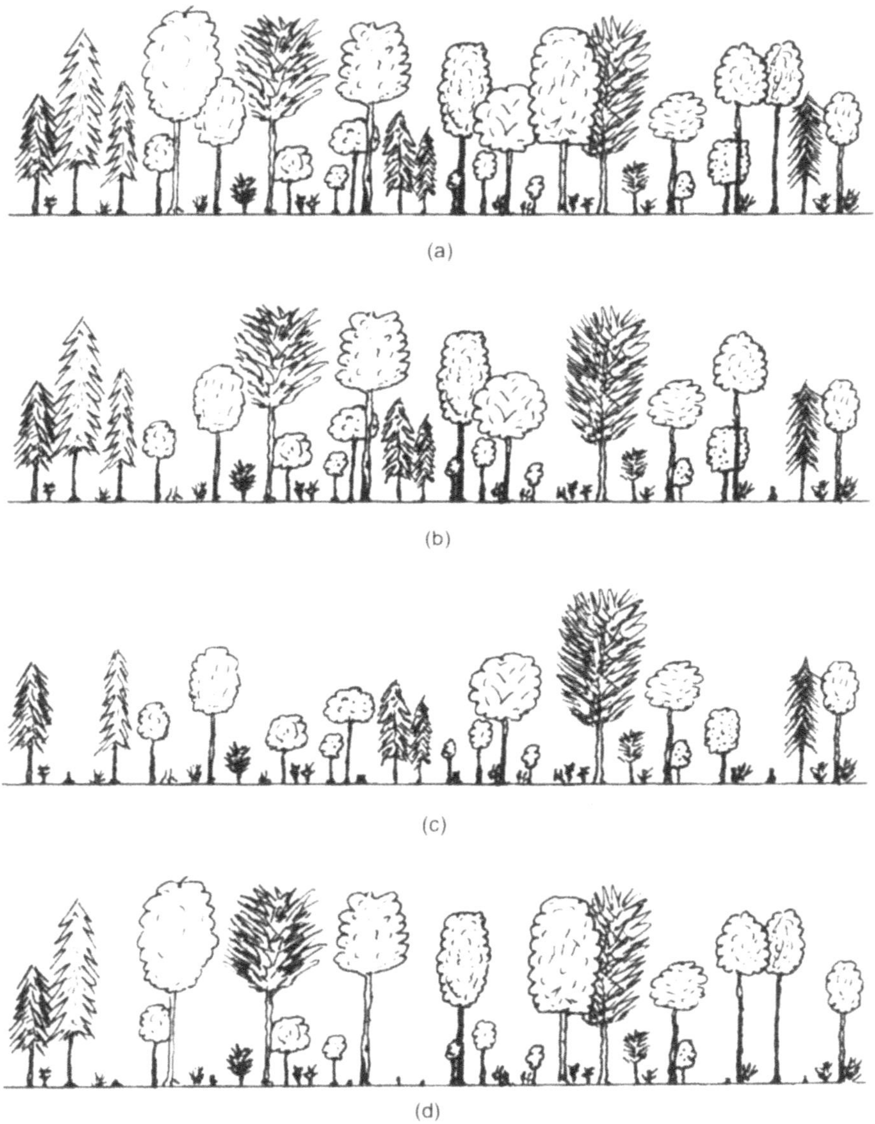

Fig. 12.2 A cross-section of a natural forest and various silvicultural treatments as discussed in the text. Illustrated are French (b), Danish (c) and German (d) prescriptions (see also Color Plate 12.1). The upper, pre-silvicultural cross-section serves as a basis for comparison.

is a larger population of trees with less of a requirement for large diameter logs.

The original French method involved removing European beech leaving various oak species (Graves, 1911; Simon, 2001). As a general prescription, this is applied to any forest where the lower value trees retard the growth of more valued species and many lesser (logs for sawmills), as opposed to few very large diameter (e.g., peeler quality) logs, are in demand.

French crown thinning is also a selective harvest method that can be carried out to remove naturally occurring latter successional trees to hasten the evolution to a climax forest. As with the climax-to-climax situation, this can also upgrade and lead to a more species and/or quality uniform forest. Figure 12.2 shows the application of a light (French) crown thinning.

Heavy crown thinning (Danish)

As a silvicultural treatment, heavy crown thinning has the goal of removing many of the dominant trees to produce fewer trees with thicker stems. Originating in Europe, the Danish method was for stands of beech (Simon, 1997), but can be applied to any situation where the objective is rapid diameter growth and superior quality in a few trees.

If done with individual trees and small gaps, the net effect will be a continuation of the climax forest. Cutting larger multi-tree gaps can favor species from earlier phases, especially in humid tropical forests with in-place early successional species. A heavy (Danish) crown thinning is illustrated in Figure 12.2.

From below thinning (German)

There is the option to keep a dense overstory and to cut the shorter intermediate trees. As with the Danish system, the goal is fewer trees with thicker stems. Although the aims are similar, the long-term outcome is different.

Developed to manage European forests, the German method starts by cutting the suppressed trees then, as a second cut some years later, the newly suppressed trees are again harvested. The first cut is shown in Figure 12.2.

Again, after some years have passed, the lesser-value intermediate and codominant trees go. What should remain are higher valued, capable of becoming large diameter, peeler-quality logs (Graves, 1911).

The French and, to a lesser degree, the Danish prescriptions produce more or less consistent and long-lasting ecosystems. In contrast, the German method has an easy to discern rotation (in terms of log size and a distinct final harvest portrayal) and, at the end, the possibility for a sudden course change.

At this point, it would be possible change to French, Danish or start anew with a from-below German sequence. This depends, in part, upon whether the large diameter trees are harvested in a light, heavy or a liberation (clear) felling.

Beyond this, other variables intercede. These include the size of the area, the in-place understory, those seed trees that stand at the ready, the nearby forest type, associated management dynamics (e.g., the application of fire), etc. The result can be a young climax forest or a jump to an earlier successional phase.

Liberation Cuts

Liberation cuts are where a large volume of overstory material is removed leading to a rapid change to another successional stage. With a large area clearcut, the result may be a high degree of stand homogeneity.

Once done with abandon, large clearcuts should be looked at through ecological lens. Rather than risk environmental consequence, gaps, cut at frequent intervals and sized to predict the successional outcome, are the alternative. Still, if a large stand is of little commercial interest and faces a bleak economic future, a radical change may be obligatory.

Within a liberation cut, not all trees need be harvested. Leaving a few of high future value (as with an uneven age plantation) can help re-seed an area with desirable species and, through regrowth and the few large remaining stems, increase the return from a future harvest.

Advancing

The common condition is to pass rapidly through a midterm successional stage to the climax forest. This is a simple process. Once the climatic understory is in place, the overstory is removed, freeing suppressed climax plants. This has been advocated in tropical situations, not as a clearcut, but as a heavy harvest (Mesquita, 2000).

Reverting

The liberation thinning may go from a more advanced successional stage to an earlier one. As an example, the Malaysian uniform system was developed to switch from mostly climax species to a high-value, light-demanding, early-phase species (Matthews, 1989; Whitmore, 1975). The requirements are large populations of prized, understory species that unsuppress when the shade-suppressing canopy is removed.

In the Malaysian case, the species *Shorea* spp., *Diperocarpus* spp., and *Dryobalanops aromatica* are desired, but some lower valued species (e.g., *Dyera costulata, Endaspermum malaccense,* and *Pentaspodon motleyi*) also occur in this successional phase and in the regeneration. The next cut, aimed at the value-offending species, is more selective and follows a French or Dutch prescription.

The Malaysian system has spawned a number of variations in tropical forests worldwide (Dawkins and Philips, 1998). Many of these

Photo 12.2 A gap in a tropical high forest. The extensive amount of biodiversity, typical of tropical ecosystems, is on display here.

involve the culling schedule for the lower value regeneration. New versions are based on gap dynamics and the resulting regeneration.

NOTEABLE VARIATIONS

The management of natural forest can have manually planted components. This can be part of an enriched natural ecosystem or an in-forest plantation. There is a progression of intensity and purpose in these variations. The intensity progression (as ordered here) is of (a) a strip harvest with natural regeneration, (b) a strip harvest enriched with natural species, on to (c) an in-forest plantation using exotics, ending with (d) a pre-forest plantation of mixed native species.

Strip Harvests

Clearcut strips can be slashed through a natural forest. Besides being able to employ strip width, through shade, seed sources, and disturbance, to encourage a continuation of the climax species, topographically sensitive placements (e.g., as with contour strips) can address environmental concerns. These can find favor in mountainous regions where a contour or similar pattern will ease the environmental (e.g., erosion) problems caused by intense logging.

Enrichments

In going beyond what nature provides in the way of regeneration, enrichment offers outcomes not available through other means. As natural regeneration often provides ample growing stock, this technique is not exercised without careful study (Magusson et al., 1999).

Even with ample regeneration, situations clearly exist where enrichment planting is the preferred option. These are: (1) where successional pressures favor less valuable species and/or (2) when the existing regeneration countenances other, less valuable, species and major change in the silvicultural prescription is not socially, economically, or ecologically feasible. For species enhancement, large openings, small gaps, and trauma-induced canopy holes can be utilized.

Even with ample regeneration, Whitman et al. (1997) in Central America found that highly desired mahogany (an early successional species) had to be planted. The gaps caused by logging were not large enough for the natural regeneration of this early successional, light demanding species, but were sufficient for enrichment purposes.

The planting method can be instrumental in achieving desired results. This is a largely unexplored topic where striplings and large stems may prove superior in survival, cost, and quality than other methods.

Stand

Forest enrichment in closed canopy stand can be part of management regime, altering the commercial value of forest, speeding growth and allowing for more frequent harvests. Enrichments can be individually placed or scattered depending on the starting point. Widely spaced trees and high ground level sunlight are one prerequisite, adequate water another.

Where natural regeneration is subject to high-level mortality, enrichments may not be economic. For this, the correct placement of individual trees (matching individual species with site) is critical, as randomly located trees may not fair well.

Gap-Strip

Gap and strip enrichment in non-uniform stands is offshoot of the management regime and harvest method. It may be better to enrich logging induced openings with an high valued species, rather than relying upon natural regeneration and realize lower-valued species (Whitman et al., 1997). Again, to be effective, age-advanced planting methods (striplings or large stems) may be needed to surmount rapidly growing, low valued, pioneer species.

Stump sprouting problems aside, strips have been advocated as a means to commercially upgrade a forest. Tagudar and Reyes (1962) mention seeding clearcut strips with the sought after species. The idea is an in-forest treatment (strip width, amount of disturbance, etc.) that duplicates those site conditions preferred by the sought after species.

Such plantings may be in clearcuts or on landings where logs were transferred from tractor to truck. Once harvest activity ceases, these become available for replanting.

Another version is to plant on temporary extraction roads. This is more a flatland exercise, as periodic harvesting in hilly and other difficult terrain may require the reuse of the same haul road network. In contrast, new skid roads on flatlands may be placed parallel to those previously constructed and later planted.

A less encountered, but promising method is to leave a few, not-to-large, standing, but dead, trees in clearings. Seed dispensing birds

Photo 12.3 A tropical in-plantation forest where natural vegetation has been allowed to naturally grow between a row of planted trees. Note the near stem weeding.

can perch on these and this helps reseed an area (Duncan and Chapman, 2003). This requires some study on habitat and effectiveness to insure that this produces the desirable regrowth.

In-forest Plantations

A variation of enrichment is the in-forest plantation. These are where plantation trees are placed within an established natural forest. Although these have the elements and the ordering of a plantation, the ecological dynamics are of a natural forest.

This can be to enrich a poor, but untouched forest or follow logging, utilizing the open spaces provided. The difference between this and other enrichment options is, for the introduced species

(a) more order and closer spacing (greater density), and

(b) continual, post-planting maintenance to insure good growth.

In an example documented by Lamb (1966), mahogany was planted in the tropical forests of Central America. For this, strips, 10 m apart, are cut in an early successional stage forest. The mahogany seedlings are line planted at five meter intervals and the strip is maintained until tree survival is assured. This introduces an ordered strip pattern into an otherwise disarrayed stand.

Exotic tree species may be added in this manner. Not being evolution-accredited members of the native ecosystem, survival and growth may be enhanced or not assured. Another possible drawback is growth interference with close-at-hand native species (Struhsaker et al., 1989).

In-plantation Forests

The in-plantation forest is a means to generate an enriched natural forest. For this, wide unplanted strips in monocultural or polycultural plantations remain. In these, natural regeneration is encouraged to proceed without interference. Photo 10.1 is indicative of this sequence where the plantation trees are taller than the natural regeneration.

This technique can be utilized to convert agricultural land to an enriched natural forest or to establish a new, highly enriched forest after severe damage (fire, clearcut, etc.). The long-term result is a native forest enriched with residual, large diameter plantation species.

Results have been mixed, some noting that aspiring understory vegetation may not match that of nearby forest in species and species mix (Loumeto and Huttel, 1997; Geldenhuys, 1997). Ashton et al. (1998) reported good results with this technique using planted strips an early successional rainforest. Harrington and Ewel (1997) found plantations of exotics not always conducive to producing a native understory. This strongly suggests that only native trees or well-studied exotics be employed if a plantation is a first step in natural forest regeneration.

Survival of the introduced trees can be insured through a brief or prolonged maintenance period. This is usually weeding around the stems. Photo 12.3 is of a in-plantation forest where strip weeding heartens the growth of the plantation species.

Survival of the natural regeneration requires (a) a wider spacing than with a pure plantation, and (b) a tree species that does not suppress understory growth. Not meeting these conditions results in a non-interference monoculture; the situation often found with a post-clearcut replanting in natural forest where a relatively pure, rather than biodiverse, stand is sought.

There can be gains letting the planted trees mature a bit where, by weeding, grazing, burning and/or a simple taungya, the onset of a scrub or a low value, pioneer forest is cancelled. Skipping the early successions and letting the planted component mature and seed can result in a latter-stage and often higher-valued successional forest.

Successional Plantations

As briefly mentioned under monocultures (Chapter 8), plantations can be a starting point for a natural succession. The plantation replaces the first successional stage, the purpose is to skip the scrub stage, accelerate the second stage, all while providing a strong economic return. The techniques for this are much the same as those employed with in-plantation forests (above), except these measures occur in the last years of a plantation. A high density initial planting coupled with thinning can trigger natural, but latter stage, successional growth.

As discussed in Chapter 8, some species are better than others at encouraging natural succession. Although exceptions exist, local species are preferred over exotics and mixed plantations of native trees seem best (Carnevale and Montagnini, 2002).

LANDSCAPE CONSIDERATIONS

The difference between a stand prescription and the full landscape can be a matter of scale. The other major differences are that landscapes cross topographical boundaries and brings into play larger landscape-wide concepts. The larger landscape, rather than individual stands, have more scale and scope to address concerns such as wildlife, water runoff and area climate.

Natural forests have less need for those auxiliary and neighboring systems that cross-confer positive ecological properties. With clearcuts and successional phases, appropriate landscape patterns can extract the most from natural dynamics. This topic is advanced further in Chapter 16.

CHAPTER 13

Agroforests

The agroforest is a major source of wood and familiar fixture in human-managed, high-rainfall, tropical landscapes. Variations exist, but are far less widespread, in dry and/or in temperate zones.

Agroforests are the agricultural version of the natural forest. These have a lot in common with natural ecosystems; both are disarrayed, ecosystem governed, and rely, to at large degree, upon natural dynamics (rather than external inputs) to achieve ecological objectives. The difference between the natural forest and agroforest is in the planning, species composition and degree of management.

Agroforests contain a mix of useful plant species with greater emphasis on fruiting and other woody plants producing non-woody outputs. These contain a high degree of biodiversity, often exceeding that of nearby natural forests.

Between agroforests and species-rich variations of their three-plus counterparts, differences exist in the amount of planning (less in agroforests, which accept considerable natural regeneration) and in governance (three-plus polycultures are mostly species-governed). The end result, agroforests are is a unique and very environmentally friendly landscape addition.

REASONS FOR USE

The important difference between natural forests and agroforests is the high productive capacity of the latter. Still, with much in common, agroforests share many of the same DAPs as natural forests. As a result, the fully functioning agroforest is the most environmentally sound form of agriculture and, because of additional biodiversity, equal on this score to the best silvicultural options.

For erosion, only the multi-storied agroforest has less comparative erosion than a natural rainforest (Wiersum, 1984). The difference is slight, but well beyond the erosion containment ability of other silvicultural systems. In other measures of sustainability, including insect and plant disease control, the agroforest excels.

Agroforests have other advantages. Wildlife can be more common in the agroforest (Perfecto et al., 1996; Cooper et al., 1996) and this can be part of a hunting strategy (see Chapter 15). As a natural fauna refuge, these systems, strategically placed, can provide pollinating insects to nearby crops and orchards (Walsh, 1997; Bawa, 1993; Gravitz, 2002).

Clearly, agroforests outshine, in all ecological measures, other planned and managed ecosystems. Although formal evaluation techniques are lacking, widespread usage indicates strong economic advantage. These may even meet the economic orientation ideal of reducing costs while increasing revenue. Not often considered a silvicultural option, these have considerable potential in farm forestry situations or for silviculture near large populations.

(See caption-next page)

Photo 13.1 Three views of tropical agroforests: (a) a closeup showing high degree of agrobiodiversity; (b) a forest garden with trees along with useful and not-so-useful understory; and (c) an agroforest located near a farm dwelling.

TYPES

Some agroforest types have been identified. These are:

(1) homegardens,
(2) shrub gardens,
(3) forest gardens.

Classification here is partially based on the output mix with corresponding purpose, intensity, species composition, and a placement which best utilizes the DAPs within the larger landscape.

For these variations, there are seldom less that 12 species and, at the high end, these can contain 200+ species. Agroforests produce a mix, but not much of one output (wood included).

Most of the silvicultural potential lies with the forest garden. The other two groupings, the homegarden and shrub garden, produce some wood but, because wood is a secondary output, these are generally not defined as silvicultural entities.

Homegardens

Homegardens intermix mix trees, treecrops, shrubs and non-woody annual species. These can contain a high percentage of or few trees, the agricultural output, rather than the wood, that is the reason for their existence.

These are often found as small stands or patches surrounding homesteads. The actual buildings forming a canopy gap, the surrounding fields producing an edge effect. As a result, these can support considerable biodiversity. In Java, 270 plant species were found across 41 homegardens and, in Mexico, 338 plants species resided in regional gardens (Cooper et al., 1996).

With this biodiversity, these provide wide-ranging outputs that are essential in subsistence and important in some non-subsistence landscapes. In league with biodiversity, Agelet et al. (2000) found that, in areas of Spain, homegardens contain 50% of the medicinal plants utilized and, in tropical regions, produce a considerable percentage of the food consumed (Cooper et al., 1996). Although secondary, 70% of the sawlogs in Bangladesh come from this source (Cooper et al., 1996). This is not because of specific purpose, but because homegardens are so widespread.

Shrub Gardens

In their permanent form, shrub gardens are usually not wood producing, except for minor amounts of firewood. The species contained are primarily short-stature perennials, the outputs are fruits and other non-wood commercial products. Shorter fruit trees (such as papaya and rose apple) and dwarf varieties are intermixed with a few taller trees and firewood species coppiced or pollarded for fast growth. These tend to be permanent, in-place fixtures.

The temporal version of the shrub garden is part of slash-and-burn sequence. This has greater silvicultural application. The sequence starts with land clearing and the planting of crops. After the crop harvest, the ecosystem evolves, entering into a planted, rather than natural, shrub phase. As the shrub garden runs its course with short duration perennials (e.g., berries, some fruits, etc.), forest trees are inter-planted. With this shrub version of taungya, the end result could be a multi-species plantation, forest garden or enriched natural forest.

Forest Gardens

The forest garden has a higher percentage of forest trees and a greater potential for wood production. Here wood can be the primary output or be an equal in value to non-woody products. As silvicultural entities, these are more akin to a highly enriched, more species diverse, natural forest and subdivide into:

(a) intense forest gardens,
(b) enriched forests,
(c) forest farming, and
(d) mixed crops and forests.

Intense forest gardens

These may be one of the more common types, mostly trees, many are fruit bearing or yielding other non-woody outputs. A breakdown of species indicates three population groupings (Salafsky, 1994: Raynor, 1992: Babu, 1992). These are:

(1) the dominant population group where a comparatively few species makeup 25-75% of the total number of plants;
(2) commonly found species that form 5-75% of the total plant numbers, these can be woody or non-woody species and makeup the largest portion of the agrobiodiversity; and

(3) the trace species where, although the total number of species is high, these constitute less that 5% of the plant total.

With considerable overlap in component species and in total biodiversity, these are best distinguished on the basis of location. Forests gardens do not abut dwellings and are mainly located on the far side of the agricultural plots. These offer considerable wood production opportunities.

Enriched forests

There are forest ecosystems that are: (1) enriched through a successional phase change, or (2) enriched climax forests. This can be forest and/or fruiting species although it is the fruiting trees that distinguish the enriched agroforest from the enriched forest.

Successional enrichment is the process of removing less valuable species as the forest grows. This leaves residual species and a species mix that would not be found in a naturally developing forest. Some internal forces encourage this. Levey (1988) found that with larger gaps and nearby fruit trees, birds spread seeds, naturally enriching and biasing plant populations in this direction.

A climax forest can be enriched by the tradition method of removing unwanted species, especially when these are still understory plants. Planting also adds forest as well as fruit-bearing species. Again, fauna can help by spreading seeds from fruiting plants.

Forest farming

In forest farming, the agroforest maintains a forest-like overstory canopy with heavily modified understory. These can be very intense, with a wide range of local and exotic species (Hsiung et al., 1995), or more limited, with mushrooms, berries and other naturally occurring native species encouraged or planted (Dix et al., 1997).

Other forest farms are more formal, with perennial shrubs are raised beneath a natural or enriched forest canopy. Cocoa, coffee and other shade-tolerant plants, those evolved for an under-canopy life, are also found. The canopy can be entirely of wood-producing tree species.

Mixed crops and forests

Also possible are merged crops and forests. These are a variation of the strip method of harvest utilized in natural forests, except the cleared strips are continually or permanently planted with agricultural crops. The non-agricultural strip can be a natural forest, a forest garden, or, more common, an enriched forest.

These forest strips serve many functions in an agricultural landscape. These can be wind or firebreak or for water management. Wang (1994) describes a tea-forest mixture where the forest strips provided climatic facilitation by controlling damaging winds and frosts. Others report less runoff and more water available to crops using alternating forest strips (Sharma and Silva, 1987).

Gap exploitation has crops planted in the openings within a forest, mixed species plantation or agroforest. Many traditional slash-and-burn system can be classified as a temporal version of a mixed gap system. Others may be less temporal, with permanent seasonal crop plots within the gaps of an agroforest (Wojtkowski, 1998). Among the advantages, this layout protects crops from frosts, winds and erosion.

MANAGEMENT

Management of the agroforest is a bit more complex than the natural forest as the value of non-woody outputs enter into the picture. Non-grazing animals (pigs, chickens, etc.) have a economic, but not a management role. Fire is generally not utilized in agroforests, these tasks are done through ecosystem governance.

The standard silvicultural treatments, those employed with natural forests (and described in the previous chapter), are difficult to implement. With more biocomplexity, multiplicity of purpose and no clear established treatments, there is a lot of reliance upon fundamental agroforest guidelines.

Fundamental Guidelines

Different management conventions exist, many depend upon user knowledge of species-based governance. A good practitioner can gain the advantages of species governance (plant-on-plant relationships) with the context of system governance, thereby living in the best of both worlds.

In natural forests, species evolve to become a full member of an ecosystem (in its fullest meaning), rather than being thrust into it (as is the case here). Agroforests are assembled from a mix of exotic and local species where new additions are subject to a trial-and-error process; one guided by a knowledgeable landuser. In this comparative environment, species thrive through slight ascendency, i.e., plants that want to dominate are demoted (pruned or planted is less favorable locations), those that do not perform well are assisted or sidelined.

Extracting the most from this, spatially and temporally, requires a profound knowledge of individual plants and their coexistence within the larger community. Lacking this for any and/or all species, fundamental guidelines come to the fore. These are, modified from Wojtkowski (1993),

(a) if a plant output is needed and the plant is producing well, leave it; if not, alter the competitive environment;
(b) if its production is not needed, neglect it;
(c) if it is negatively affecting more desirable output, prune it or as a last resort, remove it; and
(d) if space exists and essential resources are unused (as measured by the amount of light striking bare ground, plant or let something grow).

The above applies to complex unordered mix of treecrops and/or wood-producing species and to natural, but managed, ecosystems. A large part of this involves preserving or expanding biodiversity. As mentioned, better knowledge and understanding will augment or trump, or subvert these rules. Baring a more complete plan, the fundamental guidelines function well alone.

Patterns

Agroforests are light managed and canopy patterns are chosen to maximize this effect. As needed, canopies can be uniform or clumped and based on basic canopy patterns (i.e., midpoint or minimum interface). Although use is not fully understood, variations have been described (Víquez et al., 1994).

A midpoint design, with the equal distribution of light, is one option. Light can filter between the leaves of trees with open canopies or around those trees with dense canopies. This is where the 1/3 rule comes into play. This allocates 1/3 of the sunlight to each of three canopy levels. If two levels are in place, 1/2 of the light is allocated to each level. This is shown in Figure 13.1, upper cross-section.

Where there are fewer tall trees, a minimum interface pattern may be best. Here light interception is more horizontal, striking clumps from the side. The middle drawing in Figure 13.1 illustrates this pattern.

Fig. 13.1 Three disarrayed canopy patterns for multi-canopied agroforests; a midpoint (upper), minimum interface (middle), and mixed design with a upper story midpoint changes to a minimum interface closer to the ground (lower).

Some hybrid canopy designs are possible. Jensen (1993) has depicted a system where the uppermost canopy layers are in a midpoint pattern, lower down, they assume a minimum interface design. Again, this has purpose in light allocation and is pictured in Figure 13.1, lower cross-section.

With large agroforests, species composition may not be uniform, but where internal sub-stand have different outputs and purpose (Méndez et al., 2001; Gamero et al., 1996). These subsections are the agricultural equivalent of plots and, in addition to their biodiversity, but with considerable overlap, contributing additional biodiversity in the process.

SUPPLEMENTARY ADDITIONS

Given the nature of these systems, supplementary additions are far from out of place. Trace species are the norm. The more the merrier seems an appropriate guideline. Such is the case with the agroforests of the South Pacific where numerous species of shrubs, vines, herbs and ferns are found at very low planting densities (Raynor, 1992).

NOTEABLE VARIATIONS

The most common versions of the agroforest are based on full disarray. This may not always be the case and semi-disarrayed versions can and do exist. These might resemble an in-forest plantation where some tree species are ordered, others are disarrayed; others can have disarrayed overstory with an ordered understory. Given high levels of biodensity and biodiversity, these semi-ordered systems still possess superior productive and environmental properties.

LANDSCAPE CONSIDERATIONS

Positioning in the larger landscape is clearly related to type of agroforest and the categories of outputs. Homegardens are most often near dwellings, scrub gardens can be mid-way between agriculture and forestry sections. The mixed crop-forests, enriched forests and forest farming are often positioned at the farm or forest fringe.

Agroforests are not successional and rarely, if at all, do these enter a bare ground phase. With their superior ecological properties, these are assigned any number of landscape roles. These can substitute for a natural forest in intense agricultural landscapes, can replace classic farm forestry (as with forest tree plantations near crops), be at the periphery (as in Figure 16.1b), or be the center pivot (as with Figure 16.1a).

CHAPTER 14

Nature–Silvicultural Interface

A unique aspect of silviculture is the closeness to both the natural and a world defined by human activity. There are arguments for the complete protection of ecosystems (briefly presented here) and those for limited activity in forests (also briefly presented here).

What is not desired, and what too often occurs, is where ecosystems are worked to the point of destruction. Avoiding the latter, this chapter looks at the accommodation of natural flora and fauna in silvicultural context. Chapter 15 examines the second interface, that of humans in silvicultural microcosm.

Since a discussion of the world's forests must be abridged, these are divided into two categories: (1) patch and (2) gap derived. These are based on evolution of the forest type.

THE CASE FOR PROTECTION

The need to protect a forest does not require justification. History has long shown that unabated activity is not in the best interest of nature nor humankind. The section on the worst-case scenario (this chapter) delves into rapid and progressive destruction that occurs when unchecked economic forces prevail. Where conditions warrant, there can be a complete prohibition on human in-forest activities.

If silviculture is to exist, it is in the wide middle-ground where controlled in-forest activities are permitted to the benefit of people and the natural ecosystems. The argument made here encourages this view, how this is best done is an unresolved issue, one with plenty of suggestions.

Full Prohibition

The notion that good fences make good neighbors (Frost, 1915) also applies when natural ecosystems adjoin people. With a complete prohibition, no economic activity takes place and an ecosystem remains in an essentially untouched state. The possible exceptions are with low intensity, non-intrusive scientific studies or low-impact tourism.

Set aside areas are not a recent development. Alexander the Great protected an island off the Greek coast. This reportedly contained large-diameter trees and abundant wildlife (Hughes, 1994).

In the face of strong economic and social demands, natural systems do merit complete protection. Where and how much area are the key considerations. The reasons for an almost total ban on activity include (modified from Yamada, 1997):

(1) to provide a secure sample or standard that adequately represents the ecological dynamics of a specific ecosystem;
(2) to set aside
  (a) individual plants or plant communities of interest,
  (b) ecosystems of great beauty, or
  (c) areas containing outstanding or unusual characteristics;
(3) to preserve rare or endangered flora and fauna; and/or
(4) to maintain a cross-section of the genetic diversity of one or more species.

Although worth the effort, nature can have other ideas and many plants and communities may not survive in perpetuity. Plants and animals from early successions often require continual and frequent trauma (e.g., fire) as a condition of success. These plants and ecosystems can be unique in their ability to endure periodic, high impact stresses.

At the same time, climax ecosystems and their inhabitants may demand continual calm to persist. Although the timeframe may be long, many climax ecosystems do eventually encounter destructive forces. Color Plate 14.1(a) shows an old growth-forest that, in due time, succumbed to the forces of nature.

An example of enduring plants, those that have survived stress for thousands of years, are the ancient bristlecone pines of California. Those that have avoided cutting are the world's oldest living trees.

People are, over the long haul, the most destructive force and protection from human activities is often an overriding precondition for survival. The precondition is where some organization has sufficient control to ensure security. The government is a clear favorite, but non-governmental entities may also assume this responsibility. These can be corporations employing guards, religious groups overseeing sacred groves (Libbe and Freudenberger, 1996; Byers et al., 2001), estates of the rich or, more recently, conservation groups with land rights and guards.

The protection advanced should be in keeping with the strength an organization exerts. Those with more clout can prohibit more and tolerate less. Those with weaker controls or in economic need may have to permit limited exploitation.

Civil protection is at the whim of policy or political stability. History shows, protection can breakdown in the face of socioeconomic demands or national upheaval. Those islands safeguarded by Alexander the Great came to naught, but protection, if ingrained in the culture and/or religion, can survive a socioeconomic assault as long as the culture and/or religion survive.

Partial Protection

Many ecosystems cannot be fully shielded and the more pragmatic approach may be to allow limited activities. There are arguments to support this. Some have advocated limited use to ensure local people have a vested interest in keeping a forest or other ecosystem relatively intact (McNeely, 1993; Pearce, 2003a).

With this approach, the local people are entrusted with more of the protection. This is in their self-interest if there is more to be gained with sustained development than through overexploitation. Protecting the sources of irrigation or drinking water may be the most common case.

Forms of protection

Partial protection takes a number of forms, these can involve:

(1) area,
(2) harvest method (flora or fauna),
(3) species (flora or fauna),
(4) size (flora or fauna), and/or
(5) some other characteristic.

Each has advantages and disadvantages and the best option may be a combination of forms that reflect the strengths and weaknesses of the protecting body. The argument for conservation (this chapter) supposes that a set area (as with a fully protected forest) is being totally safeguarded.

Rather than complete area-based protection, an activity may be the conservation determinate. For this, people or companies are allowed to operate, but limited as to what they undertake. In silviculture, these may follow along prescribed silvicultural treatments (as in Chapter 12). For example, light crown thinning may be permitted, while other treatments are not. This second option, the harvest-management system permitted, assumes that the governing body has resources to monitor the situation. Most often, by marking the trees that will be cut, government employees insure that the permitted treatment is applied.

A third protection option is one of species. People are allowed forest activities, but may take only individual species. For example, the Government of Chile prohibits cutting the tree-species *Araucaria araucana* and *Nothofagus oblique*. Generally, there is no restriction on the type of management, if any, utilized.

As a form of control, species management works best with large trees and the need to bring these to sawmills where the offending species can be easily spotted. This control technique requires less in-forest oversight, but may not offer the degree of environmental protection envisioned. The permitted species should not be so common and ecologically key that a harvest is damaging. Usually, a lesser trace species are taken. This is more common with fauna than trees, although, with readily identifiable wood species, this is easy to implement. Due to the greater number of species, this option is better applied in humid tropical forests.

The fourth option permits species (flora and fauna) to be taken, but only within size limits. There may be minimum size and/or a maximum size. This is commonly applied to forest trees.

Most often, a minimum size will keep the stock needed for adequate regeneration, although there are doubts if this alone can be effective. Some suggest setting a minimum diameter along with limits on the species and number of trees cut per area (Sist et al., 1998). Silviculturally, this combines sparse and infrequent removals with a species-oriented sequence (Rice et al., 1997a).

A combination of maximum and minimum size can be employed to keep logging companies from conducting a 'cut and run' (quickly removing the highest value trees and then leaving the region). In a case from Liberia, an enterprise that promised to build a plywood mill was prohibited from cutting trees above a set diameter. The hope was that, once an expensive mill was constructed, there would be an economically-induced, long-term incentive for sustained yields. In the meantime, this rule preserved the larger trees, encouraging the company to keep their promise.

The fifth option looks at key characteristic and applies this to extraction or preservation. In forestry, the criteria may be a certain grade or quality of logs or be land related (as in wetland protection). These five options are not exclusive and, baring application difficulties, may work best in combination.

Outcome

The protection of flora and fauna rests on the idea that future generations will be more enlightened and appreciate past conservation efforts. Other arguments are more immediate. A link can be made between deforestation and crop failure, involving long-term reductions in rainfall and in moisture runoff that affects nearby crops (Goodland and Irwin, 1977).

The loss of a forest and the effect on water supply, both in quality and quantity, can have larger ramifications. The City of New York protects large areas of forest, permitting limited silvicultural activities, to ensure water supply and purity. The idea being is that it is cheaper to control the source rather than to pay for more elaborate water treatment. Wood sales are an economic gain that compensate for the costs of the conservation effort.

THE WORST-CASE SCENARIO

In presenting the case for silviculture, this assumes that sustainability is of paramount concern. This is not always the case. In the absence of jurisdiction, each person takes, destroying the forest and, ultimately, the land. Economists term the loss of control as "tragedy of the commons".

This is often a progression. The first harvest may be benign, but subsequent removals take more, until little of value remains. This may reach a point where, after the forest is gone, unconstrained grazing removes the last of the vegetation and erosion overwhelms the land.

There are steps along this path and, within these, the devastation may be more observable. As an example occurred with the tree species *Ceiba pentantra*. This happened in Peru where uncontrolled logging deprived an area of a key forest resource and the loss of a plywood industry (Gentry and Vasquez, 1988).

For the good, not all people and regions go the full route. A change in the social and/or economic situation or the realization that current land-use practice are destructive, can cause a course reversal. Local people have, without much outside persuasion, pulled back from the brink of total destruction (Holmgren et al., 1994; Glausiusz, 2003).

THE CASE FOR SILVICULTURE

The reasons for silviculture are many and varied, most are economic, some have an ecological justification. The most important may be in providing economic activity for local people. Protection in any form is strongest when locals have a vested interest in, and receive economic or other benefit from nearby forests. Another argument is that it is preferable to harvest, on a sustainable basis, wood from a well-monitored local forest than to import wood from far-flung, degraded forests.

The debate is not always this simple. There remain questions as to what constitutes a natural forest. Silviculture and wood extraction may not be an alien force, but can partially duplicate what occurs naturally. There is an argument that, if well formulated, constrained exploitation can promote those processes that give individual forests their unique properties.

In the regions studied, mostly northern temperate zones, one finds major changes in forest type over time. These changes span thousands, if not hundreds of thousands of years and includes most regions. Published studies exist for northeast North America (Foster and Zebryk, 1993), western North America (Lloyd and Graumlich, 1997), Scandinavia (Cowling et al., 2001), northern Spain (Penalba, 1994), and other parts of the world. The following discussion centers on the forest itself and on those influences that act or have acted on the forest and contained species as they responded to climate change and nature-induced destructive influences.

The Natural Forest

Few forested ecosystems are without catastrophic influences and few have escaped early human impact. Most have been shaped, directly or indirectly, by varying forces. Far from being incidental, forests have evolved under these conditions.

The key point is that change, not continuity, is the evolutionary norm. This fosters the view that evolutionary expectations, rather than a current state, should guide silvicultural activities (Attiwill, 1994; Coates and Burton, 1997; Wohlgmuth et al., 2002). Evidence also supports the idea that evolving, event-impacted forests are more ecological involved, maximize their internal resource use and underlying processes and, in result, produce more wood than their static counterparts (Wardle et al., 2004).

What complicates this line of discussion is that destructive influences are long term, vary in magnitude, and happen at random intervals. This makes the study difficult. Fairhead and Leach (1998) found, across time, a wide discrepancy in the recorded magnitude of West Africa forests. Inaccurate early maps and a lack of individual knowledge (now and in the past) as to what constitutes a natural forest has made prior work unreliable. With baseline data in doubt, the destructive norm is often not well defined.

Cognizance problem asides, the argument for cycles of destruction and recovery and associated evolutionary pressures is explained here using patch and gap-derived ecosystems. The evidence, and comparisons, are regional.

Patch-derived ecosystems

The natural untouched forest as an uninterrupted blanket is not an accurate picture. Given frequent assaults by strong winds, grazing animals, fire, insect infestations, ice storms, heavy snows, drought and flooding (short and long term), the median may have been patches of established forest interspersed with grassland or scrubland. The evidence for the savanna model is scattered, but compelling. Many of the world's forests, and the species contained, have evolved along this path.

Gap-derived ecosystems

Forests do overcome adversity. Rainforests clearly exhibit an evolutionary trend toward being inwardly dark, brooding places. The structure is moist, physically strong (through interlacing roots, branches, and vines), and vegetative growth is rapid enough to resist or overcome most of what nature throws at it, e.g., fire, insect infestation and all but the strongest winds.

The trauma that does occur is inner. The forest is shaped by lesser, and more numerous, tree falls. These damage a large percentage of young trees (Clark and Clark, 2001), but leave the ecosystem intact.

Tree fall gaps come about when winds or heavy rains topple large, internally-decayed trees. Gaps are the norm in tropical high forests and would be, to a lesser degree, present in and part of the traumatic evolution of all tall, dense, forest eco-types.

Brief descriptions

A complete discussion of the world's forest ecosystems is beyond the scope of this text. The following descriptions focus on the most compelling evidence and important influences, keeping in mind that the destructive forces that shape forests vary in intensity and regional importance.

Africa-The continent of Africa has interesting ecosystems, portions of which are relatively intact. The semi-arid savannas offer a window on those forces that once shaped ecosystems in other parts of the world. These are kept in a disturbed state through a diversity in grazing and other animals. The forces range from well-known mega-fauna to the less observed smaller animals.

The grazers eat up to 80% of the above ground biomass (Vera, 2000), but are not the only influence. Small, less noticed creatures can

(See caption-next page)

Photo 14.1 Additional mechanisms of forest destruction: (a) a grazed savanna, (b) trees killed by flooding (beaver induced); and (c) snow damage.

have a profound influence. Yeaton (1988) observed that porcupines, while eating bark, ring trees and, in doing so, make these susceptible to fire-induced morality.

To survive, under the pressure of fire, as well as plant eaters and plant tramplers (e.g., elephants, buffalos, wildebeest, rhinoceros, impalas, zebras and smaller animals), plants and the landscape has evolved into a form that is conducive to biodiversity and a patched layout.

Also found in Africa are well-evolved rainforests that exhibit closed-forest fauna. The presence of pygmy elephants and hippos, small deer and tree-dwelling creatures, those adopted by evolution for forest living, shows an ecosystem that has been in place, with gaps, for many millennium.

Europe-In pre-human Europe, the lowland forest, in the northern reaches and the Mediterranean, may not have been a continuous entity, but one broken up by large, open, grassy areas. As with the plains of East Africa, grazing animals once shaped the landscape. Mega-fauna such as the now extinct elephants, rhinoceros and aurocks (along with currently surviving, deer, horses, boars, bison, and beavers) certainly exerted evolutionary pressure upon the forest (Vera, 2000).

Some of the vegetation of central Europe has an evolutionary past shaped by herds of large mammals (Charles, 1997) and, in the Mediterranean region, there are tree-species suited to a savanna lifestyle (e.g., cork oak). Currently surviving stretches of savanna also attest to the past existence of this pattern (Rackham, 1998).

One must not forget fire. In the Mediterranean region, this may have been fairly common and, in northern areas, fire was less frequent, but still present. In the post-glacial forests of what is now southern Germany, major fire occurred at approximately 250-year intervals (Clark et al., 1989).

Once humans started to influence the land, the ecological situation may not have changed much. Although the mega-fauna went into decline, fire and, later the introduction of domestic animals, may have kept the dynamic state. It has been argued that ancient and still common place names, e.g., suffixes such as walde, woud, weald, and silve, denote open land, not closed forest (Vera, 2000).

Asia Those forces that shaped the ecology of ancient Europe certainly impacted the very early forests of Asia. In South Asia, some large grazers still exist and, in greater number, these would have converted a forest into a savanna (Elvin, 2004).

The long-term early evolutionary forces on Asian landscapes are less studied, but with more on later evolutionary trends. These involve the interactions of humans, useful plants, and natural organisms as they co-habit an area.

From Japan, Attenborough (2000) describes one of many situations where agriculture, forestry, and nature coexist in cross harmony with considerable mutual benefit. These mixed agricultural and forestry landscapes have been in existence for thousands of years, enough time so that minor evolution, in terms of section and informal domestication, shaped the flora and fauna.

Having attained a degree of symmetry with nature, these mixed human-forestry landscapes contain large amounts of bio- and agrobiodiversity. The existing harmony may have been initially encouraged by natural landscapes that were ecologically similar and as dynamic to those found today.

North America-Given the climatic range of North America, a number of destructive forces are at work. These varying regionally in importance and strength. Fire clearly sets the ecological balance in the west, the center exhibited both fire and grazing, while the east may have responded to mixed events.

With less diversity in grazing animals, the plains analogy might not be as accurate with the eastern forests of North America. Here infrequent, but forceful winds, may have been the predominate force. Major storms occur at the rate of five per century, hurricanes every 70-100 years (Foster, 1988).

Where powerful winds are less, fire may become the dominant force (Attiwill, 1994). The evolution of serotinus cones (those requiring fire for germination) in southern pines and some western species indicates a long fire history. Blazes can passes through an area at intervals of 2-3 years and effect individual trees every 11-39 years (Kilgore and Taylor, 1979). Conflagrations in forests destroyed by strong winds is a two-pronged ecological force, once relatively common in southeast North America (Meyers and van Lear, 1998).

In other regions, the same combination of winds and fire plays a role. The dead woody material within treefall gaps are a fire hotspot and an ecological zone, sans weeds, that favors those species that regenerate best after hot fires. This occurs with pine species in fire-shaped ecosystems (White, 1985) or with the fire regeneration requirements of the giant sequoia (Stephens et al., 1999). In many of these ecosystems, the result is an overlapping mosaic of fire-ordered patches of varying intensity.

As with early Europe, large animals made a contribution to the patch ecology of many North America ecosystems and, as with the European case, some ecosystems may have also resembled well-trodden savannas. Evidence for a highly impacted forest is shown through the native plants that once required now extinct, mega-fauna for seed spread. Honey locust, wild avocado and prickly pear fall into this category (Barlow, 2001).

The impact of human activity in pre-Colombian North America is under discussion. Some put the population levels quite high (Mann, 2002a), enough so that slash-and-burn agriculture could shape the forest (Williams, 1989).

As in early Europe, the place names used by early European settlers, e.g., the suffix field or plain, indicates that large open areas, rather than a continual closed forest, existed. It is unclear, at this stage, if the pre-Columbian forest was static or expanding (Mann, 2002a).

South America

As with other continents, South America contains many forest ecosystems. Geographically central is the Amazon region. Other parts are clearly savanna-like.

The continuous closed forest of the Amazon has evolutionary history, as indicated through the diversity of flora and fauna present, where large trees in a closed-canopy, gap-derived, forest existed. The evolution of canopy dwelling animals indicates continued evolutionary pressures in this direction.

What happened afterwards is more in question. The ability of pre-Colombian peoples to survive at high populations levels may have put this resource under pressure (Mann, 2002b; Stokstad, 2003). Similarly, the tropical forests of Panama have a 4,000-year history of human-induced disturbance, ending only after human population crashed upon European colonization (Bush and Colinvaux, 1994). How much of the forest was affected is unknown, how much of an evolutionary impact this had remains to be tested.

Natural Fauna

Clearly, local fauna have ecosystem membership and, as stewards of the forests, land users do bear responsibility for the preservation of natural fauna. This is not entirely altruistic as a diversity of fauna is instrumental in nurturing many ecosystems.

Squirrels bury acorns for future consumption and, when this process is not completed, trees grow. This is far from an isolated example, other rodents store seeds, helping regeneration (West, 1968). Seed-dispensing birds are part of a reseeding process that spans species and forests (Duncan and Chapman, 2003; Hutchins et al., 1996). One must not forget bats, Eliot (2003) described the role that bats play in spreading the tree species cecropia in the rainforests of French Guiana.

In any major ecosystem, change (even a worst-case scenario) causes some species gain, others loose. Examples across history that show gains and losses in relation to human population levels. Mann (2002a) mentions a lack of bison in the first recorded accounts of early North America, an abundance when the population of indigenous people declined, and a drop when European settlers arrived. The same holds true for the now extinct passenger pigeon (Mann, 2002a).

This also occurs within ecosystems. Studies on logged forests do show gains and losses for individual species as niches ebb or enlarge. In point, Taylor and Haseler (1995) found that, as the various canopy levels expand or shrink, so do the populations of those bird species that inhabit that level.

Management may not be this simple. Habitat is only one factor and the population respond to many variables (Anderson and Shugart, 1974).

The premise put forth is that, because the evolutionary history of local wildlife is rooted in change, ecosystem metamorphosis will have less of an impact that might be supposed (Schmiegelow et al., 1997). Therefore, maintaining a full complement of fauna species may be best accomplished employing a variety of silvicultural treatments (King and DeGraaf, 2000). Protecting or promoting any one species requires silvicultural treatments that favors the one species.

Although mostly unstudied topic, some comparatively minor silvicultural changes can produce some outwardly positive results, more so than with sweeping alterations. This can mean leaving trees that are normally cut, but are critical in supporting wildlife (Pye-Smith, 2003). In a specific study, Laiolo et al. (2003) found bird populations improved by leaving specific trees uncut and by excising more control over the species that regenerate.

Where silviculture cannot fully offer what nature expects, the option exists for enrichment, e.g., adding fruiting and other wildlife dedicated species (Lyon and Herwich, 1996). A few such trees, single or multi-purpose, will not detract from the silvicultural goals, allowing greater flexibility in selecting an appropriate silvicultural treatment.

RESPONSIBLE SILVICULTURE (BEST MANAGEMENT PRACTICES)

The argument for wood harvesting is predicated on the notion of responsible silviculture. If done in accordance with highest principles and practices, the harm done is minimal and within the parameters of natural forest disturbance.

There are some guidelines that underlie proper forestry. Those presented here summarized from Kittredge and Parker (1999).

As a protective measure, best management requires an active and effective government or landowner monitoring to approve a plan and insure that it is precisely undertaken. Those points covered here are technically applicable, with little modification, to all forest types. However, political and/or socioeconomic reality may limit what can be achieved.

Planning

A necessary prerequisite to any logging operation is planning. Good planning forestalls undue destruction of the present ecosystem while improving future productivity.

In rainforests, Nicholson (1958) observed that, in extracting an average of 11.5 trees/ha, 53% of the remaining trees above 10 cm in diameter were damaged, one-half of these severely. To put this in further perspective, Jackson et al. (2002) found that 44 trees are wounded for every one harvested. In Southeast Asia, Dauvergne (2001) established a 70% mortality rate in wounded trees. Being at the high end, this is serious destruction.

Planning can bring logging more in line with what nature expects. From established baseline data in unlogged rainforests, stem and branch breakage is common place and most smaller trees do overcome this (Clark and Clark, 2001). Still it is not good practice to drive tractors though forests looking for marketeable trees.

In Malaysia, Marn and Jonkers (1981) observed that a few simple, preplanned, control measures reduced damaged from 60 to 40 trees per ha. This need not be costly. In Latin America, Southgate (1998) found well-formulated plans increased productivity (both labor and machine use) 27%, while less organized operations wasted 26% of the felled timber. Kleine and Heuveldop (1993) noted that cost of planning is returned, even a low extraction rates.

Planning is predicated on accurate inventories (i.e., statistical samples) and/or enumerations (i.e., individual tree locations) to determine the appropriate harvest method. This includes noting those trees taken, those killed which remain standing, those cut and not removed, and those that will be the forest future. Planning is also predicated on accurate site maps that show the location of steams, wetlands, rare or endangered plants and other features of concern.

Proper Machinery

A less discussed aspect of best management is the harvest equipment and whether this is appropriate to the task. Tractors with cable winches can ease damage (Johns et al., 1996) and forwarders, rather than skidders, can reduce erosion potential.

Although clearly damaging, forest machinery may not always equate with environmental destruction. The argument has been made that the vernal pools made by wheel ruts offer numerous environmental benefits. In middle Europe, heavy equipment, through the formulation of vernal pools and the trampling of vegetation, has been shown to duplicate the effects of now extinct mega-animals; improving biodiversity and encouraging rare plants (Charles, 1997). To the contrary, one must not forget that vernal pools breed mosquitoes. For nearby, people, this has negative social consequences (Dauvergne, 2001).

Logging machinery is expensive and there is a tendency for a long in-service life. Hence, many opt for versatility and cost containment, rather than log size (length and/or diameter) and topographically specialized machinery. Within these constraints, proper planning can still be effective, even if the equipment is not specific to the chore.

Protective Measures

As part of planning and placement, there are areas that require special protection. Most involve water quality and stream preservation, but also included are wetlands, slopes and unique areas. The latter have no prescribed guidelines, management is undertaken on an individual basis. For other situations, the requirements are clear.

Road placements

Roads (haul and skid) and log landing are placed where they do the least environmental harm and where they are most efficient at extracting wood. This presupposes an accurate tree enumeration and high degree of operational planning.

Stream crossings

Stream crossings are part of the planning and placement process. Where streams or wadies are shallow with a rocky bottom, these can be forded with little damage, no special measures are taken. Where the terrain is less forgiving, other alternatives include:

(a) a poled ford (logs laid in the streambed parallel to the water flow with provisions made not to dam the stream);

(b) a reusable bridge, i.e., a temporary bridge structure that can be dragged or carried into place by tractors or other logging equipment (see color Plate 14.2b);

(c) culverts than can be temporarily placed in streams; or,

(d) if a stream is damage susceptible and machinery must be kept out of the water, a bridge is constructed from logs (see color Plate 14.2a).

Riparian areas

As a further water protection measure, a buffer zone is established on each side of an active or seasonal stream. In this zone, mechanized equipment does not operate.

Zonal protection does not rule out tree harvests, only those that taken are cut by hand and dragged or hoisted by cable to edge of the riparian zone. Cutting is done manually without tractor-mounted blades, handheld saws, including chainsaws, are acceptable. Lockaby et al. (1997) found no evidence that water quality is compromised if these measures are taken.

Wetland protection

The protection of wetlands, as with streams, is of concern and these require special operating guidelines. Wetlands are defined by vegetation and soils and include vernal pools. As with riparian zones, the wood is hand cut and cable removed.

Slope protection

Steep slopes can be subject to erosion danger and, as with wetland and riparian zones, require special safeguards. Keeping machinery off slopes is both a safety measure and an environmental necessity. Again, hand cutting and cable removal is the recommended method. Any damage done must be covered, either with logging slash or with quick growing, non-invasive covercrop.

Felling

Planning includes the felling of individual trees, both to prevent harm to loggers and nearby timber. This can be to fell downhill (this stops the tree from leaping toward a feller) and/or away from those trees that will remain for the future. The other option is a uniform felling direction that reduces damage to standing trees in the actual cutting and in subsequent skidding (Panayotou and Ashton, 1992).

When a tree topples in a rainforest, the interlacing vines can pull trees or pull branches down. Fox (1968) found that, by cutting vines and other climbers, damage to standing trees was reduced from 42% to 26%.

Slash

There are a number of options for leaving logging slash. Some areas, branches and tops of tree that remain should be cut to less than one meter above ground level. This facilitates decay, but also helps protect the ground from erosion. This is especially needed where the cut is heavy and the ground disturbed. Photo 7.1 illustrates slash for erosion control.

The burning of slash is usually not advisable except to reduce present or future fire danger. Slash does offer habitat for forest-dwelling animals and, during decay, provides an indirect food source, e.g., in the bugs contained. There may be other concerns, Wingfield and Swart (1994) found root diseases were reduced by not burning logging debris.

In some parts of the world, the tops of trees are firewood source. Moving larger tops to roads or the forest edge can be a community service and an inducement for people not to enter and cut standing trees in pursuit of firewood.

Clearcuts

Although gaps and patches can be a natural element, large clearcuts (beyond a few hectares) often run counter to what nature has in mind. There are valid reasons for clearcuts. e.g., a section of forest that is producing only low valued species and, by wiping the slate clean, the cycle starts anew, hopefully with a better species mix. Where done, clearcuts should be limited in size or, if they must be large, placed on a stable site with ecological goals in mind.

The economic forces that encourage clearcuts are not strong. Timber companies do operate profitably within more stringent guidelines. Photo 16.1 shows an erosion susceptible clearcut. This is larger than guidelines suggest.

Wildlife conservation

With silviculture, much can be done to conserve or encourage wildlife. In ecosystem management, it may be better to look first at rare and endangered species but, beyond this, questions remain as to which animals should thrive and which should merely survive.

As silviculture offers alternatives, practitioners have a lot of latitude in picking a treatment and conservation plan that is appropriate to those fauna species deemed of concern.

Those tree-species left standing is the first in a long list of options. Many practicing silviculturists, using their judgment, tend to leave fruit trees.

Within most silvicultural treatments, there is the option to kill, and leave standing, low value, poorly formed, or defect-laden trees. These rot away but, in the process, are a source of food and habitat for many birds and animals (Bures, 2005).

Trees lacking commercial value are commonly felled, but not taken. Again, these are a natural part of ecosystems and a good custom for encouraging wildlife.

Location can be important. In clear cuttings, a few trees can be left as seed trees and these also have wildlife value. Also, birds of prey, those that hunt from roost, can ply their trade from a few standing stems left in open areas.

Hollow trees are needed as dens and nesting sites for various animal and bird species. These trees, mostly lacking economic value, are best left uncut to serve as part of a wildlife-amicable plan (Mazurek and Zielinski, 2004). These are made a more attractive habitat if well located, e.g., inside a clump (Gibbons and

Lindenmayer, 1996). As a management tool and pro-wildlife measure, controlled fire increases the number of hollow trees and logs (Faunt and Williams, 1997).

In most forest ecosystems, much can be done address wildlife concerns through logging. This is a mostly unstudied aspect of silviculture, one with multiple benefits.

Abandonment

When logging or silvicultural operation ceases, some measures are taken to preserve the site. Most center on roads, but log landings and skid trails can be a worry.

Replanting

Replanting, especially in an enrichment context, is discussed under natural forest management. Clearcuts, log landings and skid trails are sites that may prove favorable. If not, these should be planted with a recommended covercrop. This can set the stage for upcoming trees or protect the site until natural regeneration takes hold. Color Plate 14.2c shows a log landing that has been seeded with grass.

Roads

A necessity in silviculture, roads are the bane of conservation and a post-harvest environmental danger point. The problem is that these serve as a conduit for illegal or unauthorized activities. Poachers in all forms find roads convenient for quick entry and escape.

Caldecott (1987) noted that, after a logging operation in Malaysia, the population of the bearded pig dropped dramatically due to hunting incursions. This is typical when unauthorized hunters have access to the once remote forest core. This can also denote the beginning of a downward spiral.

The next step can be post-silvicultural logging. A first cut may have very limited impact on the forest. This occured with an already mentioned example. For the first cut in the tropical forests of Liberia, the take was about one stem for each two to eight hectares. By itself, this was not serious. However, the roads that remained allowed more frequent, less selective cutting which, because this can be quickly undertaken and done without substantial cost, are mostly outside silvicultural guidelines and effective control.

The spiral may continue with the loss of the forest. In the Brazilian Amazon, illegal farming and land clearing is directly traced to roads. Many of which are constructed during logging operations. This problem is less documented, but equally damaging, in Africa and Asia.

In addition to planting logging and skid roads with grains or covercrops to prevent erosion, these should be blocked. In countries with strong authority-based controls, gates will suffice. Where controls are lacking, roads should be made impassable. Trees felled across roads may slow poachers, but not settlers or illegal logging operations.

For better closure, choke points should be identified and steps made to make logging roads impassable. This can mean blocking a cut through a hillside, pulling down a bridge or even blasting road breaches. These measures may seem distasteful or perhaps economically or environmentally insensitive. However, this pales in comparison when an unmonitored road starts the downward spiral that ends in a tragedy of the commons.

Inducements

Control can stem from a strong authority and/or an inducement to abide by rule, law or moral conviction (as with sacred lands). Conservation, if well thoughtout, can also be in the landusers economic interest. In any wood extraction operation, there are economic forces to maximize revenue and cut costs. These are often at odds with best management, but if properly formulated, environmentally negative economic forces may be somewhat or totally mitigated.

Take the case of logging along streams. The guidelines presented in this chapter support this, but in a controlled manner. The alternative is no extraction enforced by law or decree. If control is not absolute, this may not always be effective. Once that logging operation has ceased, and inspections made, there is a strong monetary inducement for the original loggers to conduct a 'midnight operation' or for another group to sneak in and take the marketeable trees. In this case, rudimentary environmental considerations may be completely disregarded whereas, if extracted under guidelines, this damage could be avoided or lessened.

Any harvest guidelines should be practical, not overly interfering with economic incentives. Better yet, these should be designed to encourage best practice by fostering logging methods that increase profitability. An example has already been presented (this chapter) where best management increased harvest productivity 27% (Southgate, 1998). Where preservation is and/or control is less than optimal, certification or conservation organizations should study activity logging techniques for ways to couple conservation with operational flexibility and increased profitability.

CHAPTER 15

Community Forestry

The role people and communities play, directly or indirectly, in wood production is another manifestation of silviculture. Where full protection occurs, silviculture is absent; silviculture enters the picture when exploitation takes place.

One aspect of the silviculture-community interface is in reaching an accommodation with those individuals or groups whose lives revolve around forestry activities. These can be gatherers, loggers, and wood processors that gain livelihood from the natural forest or farmers that raise trees through farm forestry or agroforestry. Forests and other wood-producing systems are also a resource for faraway communities, e.g., to moderate climate, as a source of clean water, and increasingly, for recreation.

Most germane to the discussion here is in selecting wood-production systems and harvest treatments that serve community needs the most. This should be done without overly impinging upon native flora and fauna, the character of the land, and the integrity of local ecosystems (i.e., the nature-silvicultural interface).

ECONOMIC AND NON-ECONOMIC GAINS

Whatever the landscape form, there are different gains. Some are purely economic, other activities can be harder to gauge. With monetary exchanges, community gains are easy to evaluate. More troubling to tally is the ecological destruction and environmental loss wrought by uncontrolled harvesting.

At the furthest end of the spectrum, and the most difficult to quantify, are the quality-of-life blessings. A scenic panorama; the feel of the forest after a light, cool rain; or the smell of flowers in open forest patch are nearly impossible to express in financial terms.

Benefits and relationships to local communities can be overlooked, especially when evaluating from a distance or when those with strong financial stake are the only voices heard.

DIRECT ASSOCIATIONS

The association people and communities have with trees and silviculture can be positive, both sides gaining, or negative, both sides loosing. The tragedy of the commons is one case, another is where trees are perceived as running counter to local needs. This might be where plantations have replaced traditional grazing areas. In not receiving any immediate benefit, locals may start fires to discourage forestry and return to the pre-silvicultural status quo.

Much depends upon how people view forests and trees. When trees become less of an exploitable resource and more of a means to a sustainable lifestyle, their acceptance and upkeep is assured (Holmgren et al., 1994). Examining these associations can lead to better practice.

CULTURAL SILVICULTURE

How forest and tree resources are managed is, in part, a function of local culture and, in part, imported ideas. This relates to how trees are integrated within forestry and non-forestry ecosystems and within a larger planned landscape.

Under the heading of cultural agroecology, there can be distinct landscape motif where the cultural proclivity visually imprints the land. The layout and placement of agricultural and, where appropriate, tree plots are a perceptable, if unfathomed, displayed of a culture (Wojtkowski, 2004).

Applied to silviculture, people-nature interface is outwardly manifested through slash-and-burn patterns in natural forests, religious motives that preserve some stands, or a preference for agroforestry rather than plantations. Although seldom noticed, local views can be a guide to the treatments and prescriptions acceptable to local peoples.

Community Relationships

Within cultural guidelines and needs, communities often reach a silvicultural accommodation. This can take many forms. Some clearly take without regard, others may have a long-term, less exploitive outlook.

Income opportunities are one way to overcoming a local reluctance. A thriving wood industry, with a long-term view one can ensure well-considered silviculture. More wood processing at the local level, rather than less labor-intensive chip or log exports, ties good in-forest practice with community involvement.

In poor regions where all cannot be employed, direct involvement can also encourage or aid silviculture. In-plantation grazing is an example of a community-outreach technique. This requires some flexibility in tree-species selected, planting density, and in the management directions taken. The gains can be in lessening fire danger where the reduced fuel loads and community interest results in fewer burns.

It is possible to offer local secondary products that do not detract from primary purpose of a silvicultural ecosystem. These include:

(a) guide trees, that have become unessential, useful for firewood,
(b) fruit harvested from secondary species, sometimes as a inducement to discharge some silvicultural activity, and/or
(c) logging waste carried away for individual benefit.

Product Relationships

There is a clear relationship between market demand and the types of silviculture practised. Under best management practices (previous chapter), environmentally appropriate logging equipment is discussed. This has a direct impact on silviculture practices.

Clearly, if a plywood industry is not within a reasonable transport distance, there is no market for peeler logs. This calls into question the application of silvicultural treatments designed to produce large diameter, clear logs (e.g., as with the German system). This also calls into question the need for logging equipment to harvest large diameter logs. Color Plate 15.2 is of a forwarder designed to haul small diameter logs, those destined for engineered wood products (such as chipboard or paper).

This relationship holds true for other aspects of the production process. Hauling and sawmill equipment is designed for set-range log dimensions. Few sawmills or chipmills can efficiently process a gigantic-size redwood tree. Again, this may call into question silvicultural techniques that produce logs outside set proportions, unless these become available in large numbers and mills are willing to invest in the needed machinery. Photo 15.1 has large logs being

Photo 15.1 An inefficient operation where logs are being manually sawed lengthwise to fit mill diameter requirements. Note the large, long-standing log pile in the rear, also signs of an inept, wasteful operation.

halved lengthwise to fit mill dimensions; a highly inefficient and wasteful process.

Productive Balance and Silvicultural Efficiency

The relationship between silviculture, the wood industry, and markets is seldom one of balance and efficiency. Social-economic forces can intercede to lessen the effectiveness and profitability of this relationship. This topic is seldom addressed, but examples abound and, in various forms, these are found in most regions.

Subcontracting can be efficient, but has some drawbacks. In Chile, subcontracting extends to the level of the individual worker; those felling trees provide the chainsaws, those skidding logs provide the skidders or draft animals. Since contracts are very short, payment negotiations can be problematic and companies keep large stocks of unprocessed wood on hand as a negotiating lever. The inefficiency comes as these log stocks may be better kept standing until needed, rather than degrading in storage (as in Photo 15.1).

As with labor inefficiencies, processing losses also impact silviculture. Clearly, purchasing plantation logs with little regard for quality provides no incentive for landusers to prune or make other grade improvement. Those regions that successfully produce a large percentage of high-quality plantation wood generally pay more for the best logs or pay a mill premium (i.e., additional payments made after processing based upon the grade of the lumber obtained).

A sign that the wood is being purchased without regard for grade can be piles of unusable logs outside mills. Unworkable logs are often better left standing dead in the forest or, if felled, on the ground. Both provide an ecological contribution to wildlife and the forest. This is also less costly than hauling them to sawmills where they are discarded.

In the southeastern USA, loggers mix pulp and sawmill quality logs together with the hope that these can be sold at a sawmill. The inefficiency comes when a sawmill refuses the load and, instead of being resorted, these are delivered, sawlogs included, to a pulpmill for a lesser price.

There are other disconnects. In parts of central Europe, high-quality logs from well-tended plantations are sold to small sawmills that produce and sell only low grade lumber.

Not all increases in the value of plantations and forests come through efficient silviculture. Wood wasted in processing means more must be raised and/or harvested. Although these practices may be deeply entrenched in an industrial culture, awareness and change in the post-silvicultural situation can be an environmental plus.

OTHER SILVICULTURAL-COMMUNITY LINKS

Literature places Robin Hood and his merry band in Sherwood Forest, robbing from the rich, giving to the poor. Their activities may have been part economic, part subsistence and part recreational. This may describe the relationships, poaching included, that many communities have with forests. A community, through industry, subsistence and other needs, dictates the practices employed or treatments applied. There are various elements, some more, others less influential, in the silviculture-community relationship.

Fauna Control

Natural fauna influence silvicultural and nearby agricultural ecosystems in various ways. Grazing (Chapter 6), insect control

(Chapter 7), or tree seed spread (Chapter 14) are examples of positive contributions. Effective agroecology encourages positive fauna relationships, tolerates those that are agroecologically neutral, discards or dissuades those that run counter to wood production and nearby agricultural interests. For obvious reasons (i.e., a diversity of fauna, little indication on their contribution, how the benefits are best secured), this is not a easy topic.

In some cases, the positive influences are less than apparent and any harm caused should be balanced against benefits received. MacLellan (1970) presents a case where the woodpecker damage in an orchard looked severe but, upon close examination, the benefits of the woodpeckers in insect control more than outweighed the negative aspects.

Another complication is that animals may live and feed in different eco-zones. The woodpecker example above is one such case, predator birds that control rabbits or rodents another. These offer additional complexity as well as management and control options.

In an integrated farm-forestry landscape, injurious wildlife may be best controlled where they feed. For crop-eating birds, changes in agricultural practice (feed management), rather than silvicultural practice (habitat management), may best discourage these pests. For example, some crop-eating birds do not feel safe in large fields and, to take advantage of their shyness, the sought after crops are better surrounded by less enticing fare. [Alternative feed sources may solve the problem as some birds may seek areas with weed seeds, rather that crops, if these are first presented (O'Conner and Shrubb, 1986)].

On the other side of the spectrum are those creatures that confer positive gain. Some rodent-hunting bird species can be encouraged to ply their trade outside of forests or forest fragments by placing scattered roost trees in farm fields (i.e., as with a parkland system, see Photo 1.2 (b)).

Although a component of a functioning ecosystem, fauna can be a negative influence due only to overpopulation. This is a silvicultural problem where unconstrained animal populations negatively affect commercial trees (Yokoyama et al., 2001). With a normal population, an animal can be silviculturally compatible.

Where natural fauna need to be controlled, there are a number of management options. These include (a) a less favorable habitat; (b) a habitat conducive to natural predators; (c) barriers, e.g., fences; (d) fending off unwanted animals, e.g., dogs to drive away deer; and/or (e) hunting and trapping.

Hunting

Hunting can be subsistence or recreational activity, one that endangers wildlife or, lacking a natural balance, one that keeps animal populations under control. Where wildlife can find domestic crops and livestock more delectable than what nature puts on the menu, hunting may be the easiest defense. If done well, this protects trees, nearby crops and livestock without completely eliminating a wildlife species (Giesser and Reyer, 2004).

In less enlightened form, hunting can constitute a narrowly focused tragedy of the commons. Cases exist where unregulated hunting was quite destructive across a broad range of animal species (Robinson, 1996).

In more regulated form, hunting can be a silvicultural plus. The hunting rights to the southern USA pine plantations represent a strong income source and does serve, in a minor way, to reduce undergrowth. This is a case where silvicultural practice has not been changed by the secondary activity, however, others may opt for enrichment and biodiversity to encourage sought after game.

Semi-husbandry

Semi-husbandry is where wild animals are encouraged, but not actually raised and these substitute for domestic livestock (Linares, 1976). There are advantages to semi-husbandry;

(a) less farm work,
(b) the risk of raising animals is not assumed by the landuser,
(c) the requirement for on-farm structures (including barns, pastures and fencing) is reduced, and
(d) there is less need for extended hunting trips.

In a more advanced form, those animals being encouraged are allowed to eat some of the crops or crops can be raised that will attract the sought after species. As with domestic animals, the tradeoff is in having increased protein, using some of the land area to supply this. Besides planted crops, internal enrichment (Redford et al., 1992) or trees (forest or fruit) planted around forestry blocks (Green, 1908; Peterson, 1981) can serve as feed source or enhance the habitat.

Ideally, the fauna being encouraged is not too large, not to small. Deer of various species, large birds (e.g., geese and turkeys), reptiles and amphibians (e.g., large frogs, snakes, and lizards), and small mammals (e.g., fruit bats, rabbits, tapirs, and rats) all serve this purpose in different cultures. The best animals are those do not destroy large areas of crop quickly and are easy to catch.

A form of semi-husbandry is found with the Mende of West Africa. Leach (1994) reported that rats, porcupines and small deer accounted for over 75% of the protein harvest, rare animals accounted for less than 5% of the take. In tropical Peru, 41% of the take comes from four species (Bruce, 1991).

Gathering

Many communities, not just subsistence farmers or forest dwellers, find forests a source of food (e.g., fruit, nuts, roots, etc.), fiber, fuel (e.g., firewood) and other useful materials. This can be an integral part of the diet, supplementary food, marketeable commodities, or more recreational.

Done on private or public lands, gathering need not interfere with silviculture and the community association can be quite beneficial. In Swedish forests, Kardill (1980) found that intense silviculture did not impair and may have furthered the gathering of mushrooms and berries.

Savings

In countries where banking and lending institutions are weak, income taxes are high, and/or land taxes comparatively low, trees can be a form of savings. This is where planned future harvests pay for large items (e.g., the purchase of vehicles or farm machinery) or expensive social obligations (e.g., wedding dowries, educational expenses, etc.) (Kidundo,1997; Chamber and Leach, 1990).

Saving money through tree plantings is not confined to less developed countries, but may be well established in other parts of the world. In the southern USA, blocks of pine trees may be planted upon the birth of a child, providing a college fund in 20 years. Where tax rates are high, large, unexpected income gains may be immediately invested in a tree plantation so that the money is taxed at a later date and at a low rate. In contrast, livestock, as a savings method, brings, through overstocking, environmental problems and is fraught with risk (Bradburd, 1982).

Aesthetics

The beauty of trees and forests is universal, the rich have country homes to avail themselves of this, people take day trips and vacations into rural areas to enjoy nature. Although a productive entity, this resource will not be spoiled if commercial logging is well planned and executed. With minor changes, the forestry activities can even enhance aesthetic values.

Generally, only a small segment of a forest is observable by the casual visitor. To keep logging from negatively affecting the visual experience, roads and trails can be bordered by low impact zones. These are managed much the same as riparian buffers (see Chapter 14).

The after-effects of logging can be part of an aesthetic plan. Log landings, essential to harvesting, can be positioned along roads so, when abandoned, these open small clearings that break the monotony (i.e., alley effect) of long stretches of road bordered directly by tall trees.

Along these same lines are clearcuts, not strictly for forestry purpose, but to open vistas. These are especially effective along ridgetop or hillside roads. Photo 15.2 shows a clearcut purposely placed to enhance the visual experience.

More distant silvicultural activities can also visually improved. Low intensity forestry causes little discernible disruption. However, large clear cuts can endanger and scar hillsides (as in Photo 16.1). To avoid this, harvested blocks should be small and irregularly shaped (not square or rectangular). Small, visually pleasing cuts can serve forestry purpose without economic loss while adding to the ecology and aesthetics of a region.

Aesthetics apply equally to farm forestry and agroforestry. Ornamental species, those that flower or just look nice and have no direct productive role, are often planted. This is indicative on the value placed on the garden landscape. The list of such plants is quite long and includes many useful species. A case in point, coffee shrubs are also an ornamental with nice form, foliage and fragrant flowers. Numerous tree-species also have a nice outward appearance and, even if without commercial appeal, can still adorn the fringes of a tree stand.

Photo 15.2 A vista proffered by placing a clearcut adjacent to a hillside road. The purpose is to frame the scenery for locals and increase the attractiveness of the region for tourists (photo courtesy of Massachusetts DEM).

CHAPTER 16

Silvicultural Landscapes

A forested landscape without people and with minimal human interference requires little in the way of landscape design. As a governing force, nature does just fine without outside help.

With people come silvicultural opportunities. It also goes without saying that human-populated landscapes, especially those under heavy pressure, must be overseen so that the land-use activities are mutually beneficial, sustainable and conducted in harmony with natural flora and fauna. The question lies in how this is best done.

Various factors shape human-influences landscapes. Population density, land-use intensity, culture, community values, productive needs, and outside interventions all contribute toward determining an evolving, but seldom attained, form. Of these, land-use intensity is the most visually obvious and most telling, but difficult to formally quantify (Shrair, 2000).

Some influences are less than obvious and, through these, groups and cultures put their imprint upon the land. A scan of landscapes across regions of the world shows considerable variability and, although a complete discussion of culture and other influencing factors is outside the scope of this text, presenting some basic landscape designs does help in unravelling the variations encountered.

This chapter looks at land-use intensity in landscape design. Three categories of landscapes are discussed: (1) forested; (2) fragmented, where forestry and agricultural are in separate blocks; and (3) agroforestry, where agricultural activities are either integrated or in close association with silviculture.

Other topics of interest are: (a) general land-use spatial/temporal patterns that guide many landscape layouts; and (b) changing landscape where pressures, either positive or negative, set the stage for what will come. Much of what is advocated here is the landscape equivalent of mimicry.

GAINS

Wood output is mostly a stand management issue although large-scale availability of desirable species encourages wood industry. Large areas of trees provide more than an economic contribution.

Water quality, aesthetic concerns, climatic moderation, game animals, picturesque plants and other quality-of-life additions originate at the stand level and are augmented or cancelled at the landscape (multi-stand) level. Lacking an economic incentive, these gains can fall by the wayside during economically induced change, resulting in a diminished quality of life for all concerned. With landscape planning, these can be enhanced for economic or other good.

CHANGE

Across regions, long-term trends are often the norm. This can be climatic or arise from human activity. These can be negative where trees and forests are on the wane, positive where tree populations and silviculture are expanding.

Negative

Naturally induced trauma (e.g., that from unchecked fire, ravenous mega-fauna, etc.) may have lessened in many regions, but this has been replaced by the all-too-destructive human-directed version. By consensus, a fair percentage of the world's moist tropical forests has been lost or is under threat from unconstrained logging and farming.

Where farm needs are not being met, the loss of forest often proceeds along set and predictable models. In topographically variable areas, high quality bottomland is first utilized and subsequent removal proceeds up not-so-steep hillsides. The next areas lost are the hilltops and the last to be deforested are the steeper hillsides. The process often leaves orphaned forest fragments.

Although common, this pattern is not universal. Deforestation may not be so selective and may simply proceed up hillsides, (Leach, 1994) or follow logging roads or in-forest trails.

Where appropriate, the process can depend upon ownership patterns. Block ownership without regard to topography can cause the uneven land utilization and forest fragments. Those less poor or with larger farms have latitude in not utilizing all their area.

Outside ownership and topography, the need to protect domestic animals from the hot sun causes some stands to be left. Also, the labor involved in slash-and-burn agriculture may favor reusing scrub rather than the cutting of taller forest trees (Leach, 1994). The result, blocks of highly impacted forest interspersed with more open areas.

Positive

In poor, climatically-challenged regions, the deforestation process does not always continue until all has vanished. Early losses may signal upcoming events and, having had a preview of an impending tragedy of the commons, people begin the process of rebuilding destroyed resources or ecosystems. Cases from Africa show that, although locals may be slow to adopt better practices, these come as land-use pressures increase, often invloving trees along with physical land changes (Holmgren et al., 1994; Glausiusz, 2003).

Shifting economic conditions can cause land-use modification and, where economic conditions are favorable, forests have returned after extensive human-induced clearance. In northeast USA, Foster et al. (1992) and Francis and Foster (2001) describe a landscape that changed a lot in form and appearance over a three hundred-year period. During this time, the land went through a profound cycle of clearing and regrowth. A switch from farming to manufacturing was, in large part, responsible and extensive areas of natural forest has become the current status quo.

There are other reasons for positive change. Soluri (2001) presents a case from Honduras where the closing of a railroad and a loss of distant customers changed land-use intensity. Where markets diminish, landowners often abandon low yielding farm fields. As forests naturally reestablish on abandoned and scattered farm fields, the resulting landscape may be one of farm forestry or natural forests in fragmented patches.

As land-use intensity diminishes, forest loss is not always reversed. If legal and economic conditions warrant, marginal land may be purchased by timber companies and revert to plantation forestry. This has occurred in southern Chile where residual farming is embedded among large blocks of trees.

Plantation silviculture on farms may be at the behest, but not under the ownership, of wood companies (e.g., as in parts of the southeastern USA) or as an on-farm venture (e.g., in New Zealand where farmers have embraced farm forestry). These re-establishment patterns also lead to a patch landscape.

Another restoration option, one contingent upon research and education of landowners, is that of agroforestry. As land-use intensity is reduced, taungyas replace high input agriculture. As intensity increases, rather than cultivating new land, farm intensity is augmented, not only by exploiting trees for facilitative gain, but also for their income potential (i.e., employing fully-integrated agroforestry where the value of crops and trees equate).

Reforestation, intended or unintended, partially or fully, will not always be the case as economic conditions change. Pliny mentions the giant trees of Egypt (Hughes, 1994). Long gone, these are not likely to return.

In some regions, climate may be to blame (see Chapter 6), in others, the reason can be cultural. In northern England, Scotland, and in some Mediterranean regions, locals view grazing-induced grassland or scrub, rather than forests, as the accepted norm. Although forests were once common, these are now a vestige of distant past, no longer in the memory of the culture and the people.

LANDSCAPE TYPES

Taking into account the human-induced patterns of changing land-use intensity (as above), the natural evolutionary development of local ecosystems (Chapter 14), and how the chapters on silvicultural practice (Chapters 8-13) can be implemented, three silvicultural-induced landscape forms surface. These are:

(1) the management of wide-scale plantations or natural forests with a climax or mixed successional landscape;

(2) a fragmented, mixed farm-forestry landscape that combines pastures and/or agricultural plots with silvicultural plantations and/or natural forest blocks; and

(3) an agroforestry variant, based on;

   (a) continuous, integrated agroforestry,

   (b) taungyas, or

(c) trees and crops in, small, separate, and ecologically interacting blocks.

The Forested Landscape

In some forest landscapes, agriculture is distant enough so as not to threaten the integrity of the forest. Although silviculture is practised, protection is insured through government control, private ownership or remoteness. Silvicultural treatments are chosen to maximize wood output without compromising other goals. The options are:

(a) a uniform forest treatment (with or without gaps);
(b) varying forest treatments that are:
  (i) successionally constant, but encompass different forest types; and/or
  (ii) successionally variable (as with a patched landscape); or
(c) a landscape completely covered with tree plantations.

Unvarying treatments

Where a lighter touch seems best, the ecosystem may be successionally intact. Most often, practitioners pinpoint one silvicultural treatment and apply this uniformly across the entire forest. Light crown (the French method) and heavy crown (the Danish method) may serve well; especially where the one prescription produces what the wood industry desires and where ecosystem ecology expects small gaps, but no large-scale disturbance.

Mixed treatments

As discussed, the forest need not be closed canopied. The topographical, climatic and/or vegetative differences across a large tract that make single treatment commercially and/or ecologically inefficient. The best choice may be a mix of forest types (as with the Bavarian system) where topography determines the locations and the wood requirements determine the resulting content and spatial patterns.

In an event-driven ecosystem, especially one that has been shaped by continued trauma, reintroducing the destructive influences may seem counter to productive forestry. However, as harvesting is a traumatic event, well planned extraction may fit with the evolutionary needs of the ecosystem. This includes the need for mixed successional treatments to maintain natural fauna (King and DeGraaf, 2000).

Patches are possible, either as stages of regeneration or as separate non-forest ecosystems (e.g., interspersed grasslands, wetlands, etc.). Within forest patches, gaps, either single or multi-tree, can extend the silvicultural and, by extension, the habitat possibilities. Also workable are small clearcuts or heavy shelterwoods. Mixed treatments (clearcuts included) could yield as much wood as a sustained harvest from a climax forest.

Broad plantations

A landscape may be one covered almost completely with plantations. By necessity, the blocks of trees may be in different age groups. Less than adequate from an native-plant and animal perspective are plantations of a single exotic species. The better choice is a mix of local tree-species and rotations stages, better yet are the use of multi-species plantations. Adding avenues of natural vegetation and other biodiversity features can make the landscape even more inviting for natural fauna.

Fragmented

Fragmented landscape, those with natural tree stands with interspersed agricultural plots, frame a range of land-use intensities. At one end are grazing areas mixed with blocks of natural forest. This may be ecologically similar to patches of natural grassland and natural forests (as with savannas). As land-use intensity grows, the grassland can give way to more intense cropping and/or the forest stands can become more planned and managed.

In less intense landscapes, habitat loss for natural flora and fauna can be slight. As intensity grows, there can be a loss of natural biodiversity, hopefully with gains in agrobiodiversity (e.g., with more efficient and productive multi-species plantations), although this is far from assured. The shift will be away from shy forest-breeding creatures to those that thrive in human-engendered fragmented landscapes (Donovan and Flather, 2002).

A well-formulated, fragmented landscape can still promote conservation and, through access, address community goals (Kattan and Alvarez-López, 1996; Kalkhoven, 1993; Constantine et al., 2004). Within an mixed system landscape, there will be ample opportunity to exploit the dynamics of and between individual ecosystems to allay some management concerns (e.g., fire and drought danger) and promote natural flora and fauna.

The best landscapes employ a combination of biodiversity and spatial layout to find harmony with natural forces (Wojtkowski, 2004). Some of the spatial patterns, those that apply equally to less and more biodiverse silviculture, are discussed later in this chapter.

Agroforestry

An alternative to the fragmented landscape is one of agroforestry. Here trees and crops (including forage) are temporally and spatially integrated across the landscape. While the in-field variation can be enormous, published examples are few. Three broad options are discussed:

(a) full integration where trees and crops cohabit an area and contribute in equal value,
(b) a taungya series, and
(c) farm forestry where crops and trees are separate, but in close association such that cross ecological and positive benefits of affiliation occur.

Agroforests and shade systems with a natural canopy mimic natural systems and can confer the full range of ecological benefits one expects from natural ecosystems. Similarly, well-designed and well-positioned forest or plantation stands can duplicate the habitat and dynamics of a savanna landscape. Well-placed and well-formulated taungyas do likewise.

The raising of trees integrated with and/or immediately adjacent to crops and grazing leads to both common and unexplored variants of silviculture. Color Plate 16.1c shows a mixed farm-silvicultural landscape with a small pine plantation, an orchard and integrated grazing. The small blocks promote high degree of ecological interaction between the separate systems. It is the amount of intended inter-system interaction (relative placements included) that makes this an agroforestry landscape.

CRITICAL ELEMENTS

For the three landscape categories: (1) forested, (2) fragmented, or (3) of agroforestry derivation, common elements and common patterns emerge. These enhance a landscape, not only in terms of wood output, but with regard to those responsibilities ecologists bear as stewards of the land.

Regional needs

For the complete protection of native flora, there is the question of how much undisturbed or natural, lightly utilized area needs to be set-aside. When there are few plant species, individual plots or stands can be comparatively small. Other ecosystems, especially those that are bio-rich, may need much larger areas to host a full complement of plant species. Examples are species-diverse, moist tropical ecosystems that require large traits to host a full complement of tree-species (Pitman et al., 1999; Pearce, 2003b).

The same holds true with animals. Tiny creatures survive well within small, but favorable habitats. Others require larger areas. Some percentages, high and low, have been proposed as to territory viability (Charles, 2002). Below the low number, there is not enough favorable habitat to support a breeding population. Above the high number, there is a decrease in per hectare and per species return (i.e., potential habitat may go wanting). Between these percentages, a viable population is assured.

Enrichment

The importance of plant-plant and plant-fauna associations in evolved, naturally governed, biodiverse ecosystems is beyond question. Where corridors and species movement do not suffice in keeping a species-affluent ecosystem whole, enrichment may. This is not only with wood-producing trees (as discussed in Chapter 12), but in general plant biodiversity, micro-fauna (e.g., ants and worms), and through inoculation with growth promoting micro-flora (e.g., species-specific mycorrhizae).

This can also involve multi-purpose secondary species appealing to natural fauna (fruiting trees included). These need not detract from the primary purpose of a system, as wildlife can be part of agroecological plan (as with insect-eating birds), be a quality-of-life gain (as with hunting), or be a minor annoyance that does not overly concern landusers.

Cooper et al. (1996) document an example from Java where 121 bird species are found in species-rich agroforests, 15 of which are endangered. Perfecto et al. (1996) reached similar conclusions with shade systems (crops below forest trees) and agrobiodiversity in Latin America.

Species movement

Natural corridors within a landscape, those connecting patches of natural vegetation or favorable habitats, are also part of placement. These allow both flora and fauna to move from one system to another.

Corridors or connections, either permanent or long-term temporary, prevent vegetation being lost on islands of a disappearing successional phase or forest type. Instead, these physical links allows plant species to shift to another favorable stand as one disappears.

The importance is shown by Peterken (1993), where 40% of the vascular plants found in the ancient woodlands of England did not spread to newer, detached woodlands; even if the latter had been in existence for hundreds of years. Surveys by Matlack (1994) also support this premise.

In humid tropical forest, studies show that habitat and biodiversity gains come about when adjacent gaps occur in different time periods (Levey, 1988). The best results are where a stand of one successional phase is in direct physical contact with stands of both an earlier and later phase. This includes having an early climax forest next to one that has been around for a extended period.

With the same requirements, fauna movement confirms the importance of having adjacent stands of varying ages, allowing species to migrate or pass between successional stages. Direct inter-system links allow

(a) timid wildlife to travel, avoiding contact with people;
(b) those at the lower end of the food chain a nearby closed stand in which to quickly seek a haven;
(c) tree dwellers to pass from area to area in their preferred habitat; or
(d) closed-forest living creatures to move about without being exposed to open ground and increased predation.

Forsey and Braggs (2001) found, in part of Canada, shy animals including the short-tailed weasel, red fox and snowshoe hare.

Water Management

Clean water is a by-product of environmentally friendly silviculture. This is enhanced through landscape features. Of these, riparian zones are the most critical. Located along rivers, streams, wadies, lakes, and ponds, these function as filters, removing contamination water runoff.

As part of larger landscape, these provide corridors of travel, reserves of biodiversity, serve a habitat function, and contribute to wood output. If done in accordance with best management practice, their primary task will not be compromised through the harvest of wood.

OPPORTUNE LAYOUTS

The unimpeded flow of flora and fauna across the landscape is a critical element and from this, placement guidelines can be derived. These apply equally to (1) plantations, (2) disturbed forests of varying successional phases, (3) fragmented forest and farm, and (3) agroforested. All are human disturbed and qualify as patch landscapes.

The guidelines for these have each stand bordered by a

(a) stand of a lesser and a more advanced successional phase, and

(b) by a climax forest.

Some spatial and temporal patterns are better than others at accommodating cross-landscape ecological and silvicultural needs. There are the aforementioned Swiss and Bavarian systems. If formulated in accordance with the suggested placement guidelines, these can fully accomplish both silvicultural stand and non-silvicultural landscape objectives.

Other layouts can be framed to conform to the above-stated guidelines. For example, a wedge system where each cutting takes a slice within a large circular plot (Matthews, 1989) can be ecologically boosted through intersystem associations, i.e., if bordered by blocks of climax forest. Harvesting in clusters, rather than uniformly over wide areas (Gustafson, 1996), can also help keep lesser and advanced successional stages in immediate contact.

The guidelines are more appropriate for plantations of native species. Plantations of exotics lack (a) successional patterns (but may benefit from rotations), (b) an ecological connection with native systems (but may require nearby native vegetation, as in a patch landscape, to insure good ecological things happen). With less stringent requirements, the spatial and temporal layouts for plantations of exotic trees are more akin to those of agriculture.

Gap-expecting ecosystems, e.g., humid tropical forests, should be managed more in accordance with this evolutionary requirement.

Large blocks, many square kilometers wide, can and should be internally disturbed (Pearce, 2003b). The pattern proposed here can serve as a model, but in miniature (with size appropriate gaps bringing to the fore the desired regeneration).

Relative Placement

Lacking frequent and convenient patches or strips of natural, climax forest, it is possible to design a rotational pattern that conforms to the bordering guides. There are many number of ways to do this. One such pattern, a center or ecological pivot system, is illustrated in Figure 16.1 (a). The key elements in this design are:

(a) a series of rotational blocks with, in this case, four revolving cutting cycles;
(b) the time required to complete the full cycle series;
(c) the center stands of climax forest; and
(d) a rotation sequence such that, in a given time period, like-aged stands line up to form landscape-scale strips of equal-age stands.

The size of the rotational block, the cutting cycles, and center plot are determined through wood and ex-wood criteria. The wildlife concerns are one (i.e., the already mentioned area preconditions), vegetative requirements another.

Where the center pivot is small, a key vegetative requirement is the avoidance of an edge effect in the center pivot and circling wood-yielding stands. Along the disadvantages are excessive side branching and the influx of herbaceous or weedy, light-demanding species. The edge effect can extend 15-15 m into a stand (Williams-Linera, 1990) and, keeping in mind that this happens on only the open side (per cycle), the plots should be large enough to negate this influence. With some changes in design (as in adding a buffer species), more numerous, smaller blocks might prove the better ecological option.

In some landscapes, large pivot blocks may be necessary to accommodate a full measure of diversity. There are compromises to be made. Longer rotations can shrink the size of, but not the requirement for, the pivot block.

A landscape alignment of the rotational blocks is provided for in this design (see Figure 16.2). If, for example, there is a 120-year rotation (30 years per cycle), the areas of forest or plantation align so

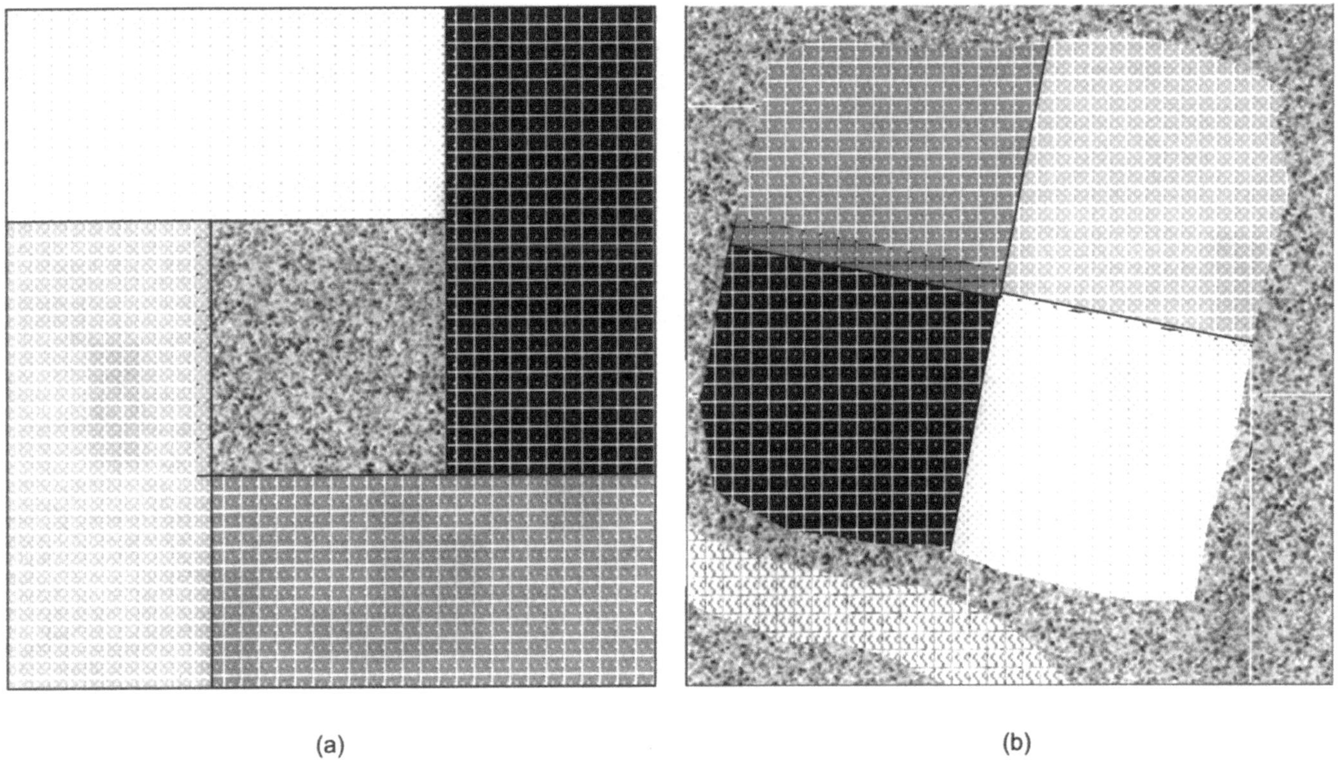

Fig. 16.1 Two landscape layouts that satisfies the guidelines proposed here. On the left is a center pivot system with a center block of natural forest. On the right is a rotational sequence with a natural ecosystem on the periphery. In both, the temporal cycles run clockwise and the age of the blocks relates to darkness (youngest = the lighter white block).

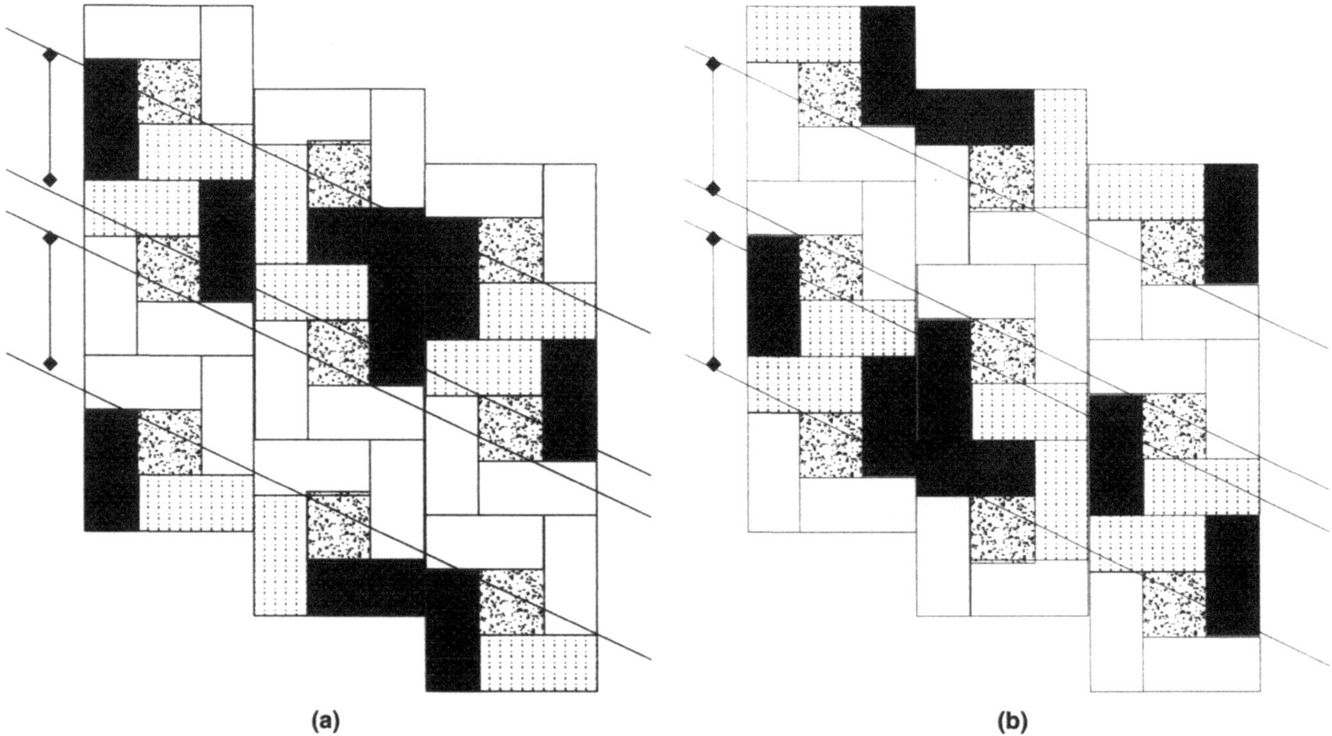

Fig. 16.2 A coordinated rotation of the center-pivot system (Figure 16.1). Through this, the blocks of like-aged trees line up to form large, cross-landscape corridors. Assuming a 30-year rotation, the two examples (a and b) are 60 years apart. The alignment is accomplished by alternating rotational direction and through relative placement of the temporal blocks.

that the natural organisms that colonize a successional phases have ample opportunity to migrate. The shift and lineup of like successional phases occurs 60-year intervals and covers vast swaths of the landscape. This is shown in Figure 16.2 a and b where the marked diagonals are alternating early successional and climax phases that abut in environmentally meaningful, long-term, cross-landscape bands.

Topographical Accommodation

The presence of corridors or clumps of climax forest (e.g., riparian zones) make the patterning process somewhat easier. The corridors can be riparian buffers or protected hillsides or ridges. Within this framework, the successional pattern is placed. Since the fringe is a climax ecosystem, the center pivot is not compulsory.

As in Figure 16.1 a topographical accommodation where streamside riparian buffers and erosion endangered hillsides provide a climax border. These should be wide enough to serve as a broad connector. The swath pattern would remain a long-term ecological force.

Photo 16.1 A large, high-risk clearcut, controversial because it goes against best management practice (Chapter 14), aesthetics (Chapter 15), and those guidelines presented here. This has the potential for natural disruption and environmental havoc (Photo courtesy of Massachusetts DEM).

Photo 16.2 The type of shy animal protected through landscape measures, in this case a North American raccoon (photo courtesy of Massachusetts DEM).

CHAPTER 17

Perspectives

In the recent past, silviculture has been viewed as wood production with few peripheral outputs or concerns. As environmental awareness has grown, this has been surpassed by a more balanced perspective.

The primary task, the growing and harvesting of wood need not negate, or even strongly compromise, environmental goals. By selecting the right silvicultural prescription, natural forests, along with forest tree plantations, can both encourage wood output and promote a range of environmental benefits. As frequently stated throughout this text, the gains include the preservation of wildlife and natural vegetation, clean water, climate moderation, recreation, etc.

With this, there are two sides to the equation: nature-silviculture and silviculture-community interfaces. With both firmly in mind, one seeks the best outcome. This can be with established techniques in plantation or natural forest management or, where convention is not the right course, alternative techniques and agroecological tools can be brought to bear.

Natural ecosystem mimicry, including natural cycles of forest destruction, help make productive silviculture into something that nature approves of. This applies to plantations, with a beginning and an end, and to natural forests where ruinous forces often rule.

Less utilized techniques, such as multi-aged plantations and in-forest plantings, transcend traditional subdivisions, adding biodiversity where intended. Other tools are more localized, hands-on, and involve what to do in certain situations. Inclusive are the spatial and temporal patterns and the planting and weeding options.

There is much to be gained going outside conventional practice. Since only scattered studies or accounts underlie the bulk of non-mainstream silviculture, efforts along these lines are based more on good judgment rather than scientific clarity. Accordingly, silviculture, despite the large body of research, is still more an art than a science.

SCOPE

Yields and environmental benefits can accrue in managed forests, dedicated plantations, and in agricultural settings. The latter brings about an expansion of traditional silviculture.

This is not new. Shadowing silviculture, but at a respectful distance, has been agroforestry. Some of the first silvicultural texts mention agroforestry (Gent, 1681; Browne, 1832) as do the more or less recent books (Schenck, 1904; Evans, 1992). As agroforestry has advanced, an agriculture-forestry split has become apparent. Wood-first agroecosystems are inclusive under the heading of silviculture and, by extension, forestry. Agroforestry is part of these as long as wood is the primary output. This does do not encompass all of agroforestry. Outside these bounds are those agroforestry systems where wood output is minor to crops.

The logical partition with the agriculture-forestry dichotomy is to roughly include agroforestry where wood production is a primary or a joint purpose; other systems lie elsewhere. However, if one includes those systems where wood production is secondary to agricultural output, silviculture risks coming out from under the forestry definitional umbrella.

VARIATIONS

The methodology chapters in this text (those on the various silvicultural practices) are roughly ordered by the degree of biocomplexity. A genetically pure monoculture is the first presented, ending with highly complex agroforests. Placing monocultures first does not suggest a superiority of bio-simplicity. Quite the contrary, bio-complexity is a powerful ecological and economic tool and may be preferable. Nevertheless, monocultures are basis of comparison and this sequence should be presentationally revealing.

The silvicultural options are quite extensive, many have remained unnoticed and unexplored. There are reasons from going beyond the current norm; spatial patterns administered through biodiversity can offer big rewards in terms of yields, profits, cost and risk reductions and environmental gains. Other avenues for improvement lie in temporal plane with or without biodiversity (e.g., multi-age monocultures or bicultures).

ECOSYSTEM TYPES

Within silviculture, a number of subdivisions occur. Well established are the difference between (a) tropical and temperate, and (b) humid and dryland systems. These are mostly dictated by the site.

Less geographically pronounced, but still descriptive, are systems that are: (c) species and ecosystem governed, (d) ordered and disarrayed, (e) naturally occurring and fully-formulated, (i.e., planned), and (f) patch and gap-derived.

The point being made is that multiple descriptors lead to more precise understanding, better diagnostics, better treatments and better outcomes. That is, besides defining and describing, these aid in changing the silvicultural, ecological, natural, and/or economic focus on a given site, with given species, and with established goals.

Tropical and Temperate

Most differentiate between moist tropical and temperate silviculture. The major difference is in the amount of biodiversity and resulting management complexity.

There are some comparative figures, temperate New England has about 5,000 plant species whereas moist tropical forests contain up to 30,000 species (Withner, 1977). For trees, 300 species are mentioned in connection with the Peruvian rainforest (Rice et al., 1997b) while others report fewer, e.g., the 132 species found in parts of the Caribbean (Forman and Hahn, 1980).

Bio-complexity extends to fauna. Fleming (1973) found 15-16 species of mammals in Alaska, 31-35 on the east coast of North America, and around 70 in the moist forests of Panama. Likewise, the tally of insect and bird species in untouched rainforests is exponentially greater when compared against temperate counts (Rice et al., 1997b). Again, this is a complicating factor, but with an acceptance and understanding of silviculture based upon biodiversity, these differences pale.

Humid and Dryland

Water, as the limiting essential resource, goes a long way toward defining the humid-dryland dichotomy. Whether tropical or temperate, the dryland situation demands water management. The preoccupation with tree survival and the specialized knowledge required drives this version of silviculture.

In contrast, humid areas are free from this constraint. High levels of biodiversity and associated concerns remain an obstacle where moisture is abundant.

For good reason many of the silvicultural treatments have European place names. European forests have comparatively fewer tree species. This lends a certain clarity to the process whereas, in the moist tropics, the diversity, the lack of usable growth rings, and the large distances between like species (Pitman et al., 1999; Sterner et al., 1986) makes silviculture more problematic.

Still, the basic silvicultural prescriptions (e.g., the French, German, and Danish treatments) have application to drier situations and for guiding efforts in both humid temperate and humid tropical forests. To be fully applicable, they must undergo some modification. As with temperate trees, tropical species can be divided into temporal groupings, e.g., pioneer, early successional, etc. One difference is that, for humid tropical forests, there are far more species in each temporal class.

Species and Ecosystem Governed

A group of plants can be governed by the interactions of a few key plant species or by the collective will of many organisms. Where the influential ecological dynamics and productive (wood) gains originate from a few plant species, this is species governance. With many organisms (micro and macro) contributing, a ecosystem becomes more than the sum of the parts. This is ecosystem governance.

Dividing ecosystems along governance lines brings out that

(a) stand dynamics originate at various levels;

(b) inappropriate dynamics can be counterproductive, e.g., too much biodiversity overwhelms the key species, messing up growth rates;

(c) an appropriate form of governance can do much to advance productive and natural objectives; and

(d) once an ecosystem type and path is selected, there are design parameters and management methods for each.

Chapters 8, 9, 10, and 11 mostly discuss species-governed ecosystems whereas Chapters 12, 13, and 14 present the parameters and techniques that underlie ecosystem governance.

As an economic entity and/or a landscape component, a stand can be viewed as a group of trees (species governed with DPCs) or as an ecosystem (ecosystem governed with DAPs). There are inherent dangers in taking an unstudied a view of these subdivisions.

Species governed systems (e.g., plantations) are not artificial constructs outside the purview of nature, i.e., these should be viewed less as a collection of useful plants, more as an ecologically-forceful contributor to natural ecology. Irrespective of governance, all systems have DAPs and these appraise how well planned and management ecosystems foster natural flora and fauna and partake in the natural ecology of the surrounding landscape.

Ordered and Disarrayed

This text has made a clear distinction between ordered and disarrayed systems. Again, this goes to precondition. Ordered systems are often associated with planning (as with plantations). However, there are exceptions. Disarray can be a planting option for a multi-species plantation. Agroforests, planned and often planted, thrive on disarray.

This dichotomy should not be associated with any one form of silviculture. Besides being a category, this can also be considered a management tool, e.g., utilized to overcome knowledge limitations in dealing with agrobiodiversity (see Chapter 3).

Natural and Planned

The natural and planned dichotomy, in contrast to that of order and disarray, often separates systems derived by people from those put forth by nature. In most cases, this is fairly clear; plantations fall in one category, natural forests in another; one is ordered, the other disarrayed, one is species governed, the other ecosystem governed.

These groupings are not always precise. Agroforests are planned, but disarrayed and ecosystem governed. This combination couples high productivity with environmentally friendliness.

In-forest plantations (a tree plantation placed in established forest) or in-plantation forests (allowing a natural forest to grow within an established plantation) also transcends what can be a simple divergence. These are semi-ordered, somewhat planned, ecosystem governed, can incorporate counter-successional native species or have exotic species thrust into the mix.

The lesson here, as with other cross groupings, is that categorical purity may not be the best choice. Crossing conceptional boundaries, as with natural and planned systems, can offer productive and environmental advantages not possible staying within a single ecosystem type.

Patch and Gap

The patch and gap subdivisions derive from the underlying evolutionary patterns of most forests. Although exceptions exist, gap ecosystems are mostly found in moist tropical high forests, patch forests are more a product of temperate and/or semi-arid landscapes.

Gap forests are capable of maintaining a closed canopy through copious growth. The gaps result from very large trees, some tied together with vines, which, upon falling, tear holes in the canopy. Despite the fact that these canopy openings are comparatively shortlived, they continue to appear. As a result, the gap ecosystem is dominant.

In contrast, patches are found in those ecosystems that have difficulty overcoming the forces of nature. These forces, grazing, tree-eating fauna, high winds, fire, heavy snows, flooding, etc., all strive, alone or in concert, to unmake the forest as a single boundless entity. A mix of forest stands and non-forest ecosystems results, the magnitude of each traumatic event dictates the mix.

The patched savannas of Africa are remnants of what was once a widespread land type. In some regions (e.g., large parts of Europe and the Americas), the elimination of unchecked fire and large herds of grazers have caused forests, those that still remain, to lapse into a closed canopy form. Despite being less visible nowadays, the dynamics and component species are still part of the evolutionary underpinnings. Many species of flora and fauna benefit if the original dynamic order is returned and maintained.

The implications are in

(a) the ecosystem, i.e., the flora and fauna, that are expected on any given site;
(b) the amount and type of trauma which these can endure;
(c) the human-induced activities (fire, logging, etc.) which replicates what the ecosystem expects.

Duplicating exactly what nature does is a good starting point. Where this is not possible nor productive, appropriate silvicultural treatments, those that follow nature's prescriptions, can be a surrogate to bring the land back to the underlying and expected evolutionary pattern and on to optimal ecological health.

AGROECOLOGICAL APPROACHES

Whether a purely climatic category, (e.g., tropical and temperate or humid and dryland) or those with broader application, an appropriate prescription helps set the stage for sustainable silviculture. The choice between order and disarray, patch and gap and natural and planned all denote an approach.

Other concepts also underlie silvicultural efforts. Not all can be summarized, but a few points, those that may have been partially obscured amongst the practice details and subject cross threads, merit further comment.

Biodiversity

A hallmark of agroecology is the reliance on biodiversity. With a directed version, plants are incorporated because they have productive output or they can facilitate the growth of those that do. With agrobiodiversity, all plants have productive purpose and multiple roles in contributing to the ecological whole.

Not be overlooked is incidental biodiversity, those species, plant or animal, that arrive unannounced. This is not always bad.

If species governed, there is an attempt to imbue all niches and assign all ecological roles. If system governed, enough biodiversity (micro and macro, flora and fauna, etc.) should be present so that all niches are filled, all roles taken up. The key is being able to focus biodiversity for productive and, as a corollary, for environmental good.

Design

The concept of design is one that consolidates multi-use forestry. Agroecosystems (silviculture included) are specifically formulated to produce economically interesting output. It is well recognized that these same ecosystems have other properties, i.e., the DAPs. These may include an ability to control erosion, minimize damage from herbivore insects, deflect strong and damaging winds, counter drought, or deal with other on-site hurdles and pending calamities.

Most are content to select an appropriate practice from a set, biodiversity-systematized list (such as those presented in the practice chapters). The more adventuresome look elsewhere and design their own systems. Those unique, one-of-a-kind practices that illustrate major divisions of biodiversity originate in this way; through design.

Design is the process of formulating ecosystems for purpose, DAPs included. To bring to the fore the DAPs, the species mix, planting densities, secondary vegetation, spatial and temporal patterns, along with management inputs are employed.

For example, the adding of second species, if well singled out, should result in a large ecological boost, a change in DAPs, and a shift in economic direction. Knowing this, one objective can be to replace costly inputs with natural forces (e.g., replacing insecticides with predator insects). Other goals can be handled in much the same way. A second species may be enlisted to promote those natural dynamics that substitute for manually performed tasks (such as weeding, thinning, and/or pruning).

Ideally, a correctly formulated stand can both reduce costs and risks while increasing yields. In silviculture, this is possible, especially if all the forces that drive natural ecosystems can be focused and brought fully to bare.

The number of options and directions are enormous. Most circumvent these by selecting a prestudied suitably-tried system and appending ecologically and economically contributory features (e.g., a different spatial or temporal pattern, directed biodiversity, management tools or options, etc.).

However, it is done, this is an inexact science that, once mastered, offers substantial silvicultural potential. As with other concepts, design is not exclusive, but is useful for the options offered.

Mimicry

The application of mimicry offers a rationale for silvicultural intervention and a view towards what may work best. The idea being that those ecosystems which are ecological appropriate to a region and site are best for that region and site. Ecologically appropriate means ecosystems that resemble (in ecological character, function, dynamics, and, to a lesser degree, species composition) those naturally found.

A well-managed natural forest that maintains the biodiversity and dynamics of an untouched natural forest may be highly appropriate to yield requirements and ecological needs. Where more intrusive wood producing enterprises are needed, a highly biodiverse plantation might carry on the ecological tradition associated with a site.

Mimicry does not exclude monocultures and bicultures. These are found in nature and, if part of a well-planned landscape mosaic, these can be an appropriate land use.

Mimicry is an extension of the ecosystem types where one selects and plans based on (a) governance, (b) formulation (i.e., natural or planned), (c) order or disarray, and (d) patch or gap-derived. Management also comes into play in an attempt to duplicate routine, but often stochastic, traumatic events. For the latter, gap size, frequency of patches, intensity of a burn, types of animal grazed, etc. can inaugurate and/or maintain what nature has come to expect.

POLICY

Land-use policy is a difficult topic as forests and those forces, destructive or otherwise, that influence their development are undefined. Also, public opinion does not always accept what is right, if indeed, this is known. Take the case of three-plus polycultures. Once sanctioned and then actively scorned, these fall within the precepts of biodiversity and are recognized for their productive and ecological potential.

Other issues have not been fully addressed. Many regions have overlooked agroforestry as a land-use option and, in the process, have negated potential gains in sustainability and productivity. A case in point, in developed regions, over production in agriculture and the loss of farm income can be rectified by turning to plantations that mix crops with trees. Once surmountable obstacles, including policy recognition, are overcome, farmers benefit with higher incomes and an environmentally friendly landscape.

Other aspects of policy look at the natural forest destruction, mainly fire, but other forms also qualify. Often viewed as environmentally adverse, a strong argument can be made that natural destruction, where present in the evolutionary past, should be a component undertaken at opportune junctures, in active forest management.

A lack of a clear agroecological direction and a knowledge of the full complement of alternatives limits what can be accomplished in formulating environmentally friendly land-use practices. Carried across to policy, there is a risk in limiting the silvicultural options and therefore the productive and environmental gains.

RESPONSIBILITY

Beyond formal policy is inward responsibility. Stewards of the land have the obligation to respect natural flora and fauna and to keep the land productive into the future. Without agroecological principles and practices, there can be a tendency to compromise. However, if the full range of land-use possibilities and management options are exercised, environmental and productive compromises are far less needed. The ideal, seeming beyond reach in productive setting, is far more attainable.

CONCLUSION

The human experience is clearly improved with trees. Firewood and materials for construction are often in short supply, a product of silviculture, and an economic inducement for planting trees.

In intact forest ecosystems, serviceable wood can be foremost on a long list of output and benefits. Additionally, forests are also a source of native plants that provide foods, medicines, and other non-timber commodities. Clean water is yet another blessing while wildlife, indulged through sympathetic silviculture, can offer locals protein and/or tourist revenue. Quality-of-life gains, including the beauty of forests and trees, also better the human experience. In short, silviculture improves livelihoods and landscapes.

Outside human needs, responsible silviculture can be a platform to support natural flora and fauna. This can be within specific ecosystems or as a separate refuge or haven in an otherwise inhospitable landscape. Wood producing ecosystems such as forest-tree plantations, agroforestry systems, or natural forest fragments, can provide a safe harbor in an otherwise unfriendly agricultural landspace.

An often uneasy relationship exists between silviculture and environmental science. This should not be the case. As an accessory to ecology and environmental science, a rich silvicultural pallet can provide ecologically wholesome interventions; favorably realized if part of an economically positive picture. Wood production, through a focus on the growing, harvesting, and selling of trees, is the driving engine.

Silviculture, need not, nor should it be the enemy of good land-use practice. Much more can be done to avoid land-use pitfalls and in having a landscape in conflict with nature. In this age of ecology, this is best accomplished through a revised and agroecologically-mobilized silviculture.

Begin in time to thin your trees
And think not you have some to please
But let each tree has space enough
And soon you'll find this is no stuff;

But sound and clear's the truth I teach
if it could to the mind but reach
And be fixed there, both firm and fast;
Then I should have my wish at last.
(Smith, 1870, p.92)

Color Plate 4.1 a and b Successional phases: (a) the intermediate point in an early successional phase showing residual scrub vegetation, (b) a conifer stand being temporally overtaken by a climax hardwood forest.

Color Plate 12.1 Three removal options: (a) a light crown (French) thinning, (b) a heavy crov (Danish) thinning, and (c) a thinning from below (German).

olor Plate 14.1 Mechanisms of natural forest degeneration: (a) wind destruction in a mature and; (b) insect defoliation; (c) a small localized fire (Courtesy of Massachusetts DEM, the fire photo as taken by Tim Zelazo).

Color Plate 14.2 Good harvest practice: (a) a log bridge to protect a stream; (b) a temporary ro cover to prevent erosion, and (c) planted logging road also to reduce erosion (Courtesy Massachusetts DEM).

Color Plate 15.1 A local gathering discarded treetops for firewood. This is part of a free cleanup before the replanting of an intense poplar plantation (as in the background).

Color Plate 15.2 Two logging operations; (a) A forwarder hauling pulpwood, this is less damaging, but limited to smaller-sized logs and (b) larger logs that have been skidded to a road.

Color Plate 16.1 Different forms of fragmentation: (a) an open landscape with distant tree blocks; (b) farmer encroachment up a forested hillside, and (c) farm forestry with a pine stand integrated into the overall system.

Color Plate 17.1 The silvicultural ideal, clear, straight, high-value trees growing in ecologically healthy, well balanced, and

References

Agelet, A., Bonet, M.A. and Vallès, J. 2000 Homegardens and their role as a main source of medicinal plants in mountain regions of Catalonia (Iberian Peninsula). Economic Botany 54(3): 295-309.

Altieri, M.A. and Nicholls, C.I. 2004 Biodiversity and Pest Management in Agroecosystems. 2nd edition, Food Products Press: NY, USA. 236p.

Altieri, M.A. 1994 Biodiversity and Pest Management in Agroecology. Food Products Press: NY, USA. 185p.

Anderson, L., Carlson, C.E. and Wakimoto, R.H. 1987 Forest fire frequency and western spruce budworm outbreaks in western Montana. Forest Ecology and Management 22: 251-260.

Anderson, S.H. and Shugart, H.H. 1974 Habitat selection of breeding birds in an east Tennessee deciduous forest. Ecology 55: 828-837.

Anon. 2002 Bee off with you. The Economist 365 (2 November - 8297): 78.

Ashton, P.M.S., Gamoze, S., Gunatilleke, I.A.U.N. and Gunatilleke, C.U.S. 1998 Using Caribbean pine to establish a mixed plantation: testing effects of pine canopy removal on planting of rainforest tree-species. Forest Ecology and Management 106: 211-222.

Ataroff, M. and Rada, F. 2000 Deforestation impact on water dynamics in a Venezuelan Andean cloud forest. Ambio 29(7): 440-444.

Attenborough, D., Hogarth, J. and Saito, S. 2000 Japan's Secret Garden. NOVA Broadcast No. 2716, PBS air date: December 19, 2000.

Attiwill, P.M. 1994 The disturbance of forest ecosystems: the ecological basis for conservative management. Forest Ecology and Management 63: 247-300.

Babu, K.S., Jose, D. and Gokulapalan, C. 1992 Species diversity in a Kerala home garden. Agroforestry Today 4(3): 15.

Bach, C.E. 1988 Effects of host plant patch size on herbivore density patterns. Ecology 69(4): 1090-1102.

Bainbridge, D.A. 1985 The rise of agriculture: a new perspective. Ambio 14(3): 148-151.

Bainbridge, D.A. 1988 The oaks: a neglected multiuse treecrop. pp. 657-601. In: P. Allen and D. van Dussen (eds.) Global Perspectives on Agroecology and

Sustainable Agricultural Systems: Proceedings of the Sixth International Scientific Conference of the International Federation of Organic Agriculture Movements. The Ecology Program: Santa Cruz, CA, USA. 730p.

Baker, F.S. 1934 Theory and Practice of Silviculture. McGraw-Hill: NY, USA. 502p.

Banda, A.Z., Meghembe, J.A., Ngugi, D.N. and Chome, V.A. 1994 Effect of intercropping maize and closely spaced *Leucaena* hedgerows on soil conservation and maize yield on a steep slope at Ntcheu, Malawi. Agroforestry Systems 27: 12-22.

Barlow, C. 2001 Ghost stories from the ice age. Natural History 110(7): 62-67.

Barton, A.M. 1999 Pines versus oaks: effects of fire on the composition of Madrean forest in Arizona. Forest Ecology and Management 120: 143-156.

Baskin, Y. 1994 Ecologists dare to ask: how much does diversity matter? Science 264 (8 April - 5156): 202-203.

Bawa, K.S. 1993 Plant reproductive biology and agroforestry. pp. 95-99. In: W.R. Bentley, P.K. Khosla and K. Seckler (eds.) Agroforestry in South Asia. Winrock International: USA and Oxford and IBH Publishing Co.: New Delhi, India. 367p.

Bellefontaine, R., Gaston, A. and Petrucci, Y. 2000 Management of Natural Forests of Dry Tropical Zones. FAO Conservation Guide, FAO: Rome; Italy. 318p.

Benson, L. 1977 Preservation of cacti and management of the ecosystems. pp. 283-313. In: G.T. Prance and T.S. Elias (eds.) Extinction is Forever. New York Botanical Garden: NY, USA. 437p.

Bezkorowjanyj, P.G., Gordon, A.M. and McBride, R.A. 1993 The effect of cattle foot traffic on soil compaction in a silvo-pastoral system. Agroforestry Systems 20: 1-10.

Blanford, H.R. 1925 Regeneration with the assistance of taungya in Burma. The Indian Forest Records 11(3): 1-39.

Borthwick, A.W. and MacDonald, G.V. 1913 Nursery and plant competition. pp. 85-101. In: A.W. Borthwick (ed.) Transactions of the Royal Scottish Arboriculture Society, Vol. 27: Edinburgh, UK. 248p.

Boutcher, W. 1775 A Treatise of Forest Trees. R. Fleming: Edinburgh, UK. 259p.

Bradburd, D. 1982 Volatility of animal wealth among southwest Asian pastoralists. Human Ecology 10(1): 85-106.

Bragança, M., De Souza, O. and Zanuncio, J.C. 1998 Environmental heterogeneity as a strategy for pest management in *Eucalyptus* plantations. Forest Ecology and Management 102: 9-12.

Brenner, A.J. 1996 Microclimatic modifications in agroforestry. pp. 159-188. In: C.K. Ong and P. Huxley (eds.) Tree-Crop Interactions: A Physiological Approach. CAB International: Wallingford, UK, 386p.

Brose, P., van Lear, D. and Cooper, R. 1999 Using shelterwood harvests and prescribed fire to regenerate oak stands on productive upland sites. Forest Ecology and Management 113: 125-141.

Brown, J. 1861 The Forester or A Practical Treatise on the Planting, Rearing, and General Management of Forest Trees. William Blackwood and Sons: London, UK. 700p.

Brown, J.C. 1878 Pine Plantations on the Sand Waste of France. Oliver and Boyd: Edinburgh, UK. 172p.

Browne, P.J. 1832 The Silva Americana. William Hyde and Co.: Boston, USA. 407p.

Brown, J. 1861 The Forester of A Practical Treatise on the Planting, Rearing, and General Management of Forest Trees. William Blackwood and Sons: London, UK. 700p.

Brown, J.C. 1878 Pine Plantations on the Sand Waste of France. Oliver and Boyd: Edinburgh, UK. 172p.

Bruce, J.W. (1991) Securidad Alimentaria Familiary Silvicultura: Análisis de los Problemas Socioeconómicos. FAO, Rome, 150p.

Bryant, A. 1871 Forest Trees for Shelter, Ornament and Profit: A Practical Manual for their Cultivation and Propagation. Henry Williams: NY, USA. 244p.

Budelman, A. 1988 The performance of leaf mulches of *Leucaena leucocephala, Flemingia macrophylla* and *Gliricidia septium* in weed control. Agroforestry Systems 6: 137-146.

Bures, F. 2005 Dead wood talking. Audubon 107 (March–2):18.

Buresh, R.J. 1999 Phosphorus management in tropical agroforestry: current knowledge and research challenges. Agroforestry Forum 9(4): 61-65.

Bush, M.B. and Colinvaux, P.A. 1994 Tropical forest disturbance: paleocological records from Darien, Panama. Ecology 75(6): 1761-1768.

Büttner, V. and Leuschner, C. 1994 Spatial and temporal patterns of fine root abundance in a mixed oak-beech forest. Forest Ecology and Management 70: 11-21.

Byers, B.A., Cunliffe, R.N. and Hudak, A.T. 2001 Linking the conservation of culture and nature: a case study of sacred forests in Zimbabwe. Human Ecology 29(2): 187-218.

Caldecott, J. 1987 Hunting and Wildlife Management in Sarawak. World Wildlife Fund: Malaysia (referenced in Panayotou and Ashton, 1992).

Callaway, R.M. 1992 Effects of shrubs on recruitment of *Quercus douglasii* and *Quercus lobata* in California. Ecology 73(6): 2118-2128.

Cappuccino, N., Lavertu, D., Bergeron, Y. and Régniėe, J. 1998 Spruce budworm impact, abundance and parasitism rate in a patchy landscape. Oecologia 1142(2): 236-242.

Carnevale, N.J. and Montagnini, F. 2002 Facilitating regeneration of secondary forests with the use of mixed and pure plantations of indigenous tree-species. Forest Ecology and Management 163(1-3): 217-227.

Carucci, R. 2000 Trees outside the forest: an essential tool for desertification control in the Sahel. Unasylva 51: 18-24.

Chambers, R. and Leach, M. 1990 Trees as savings and security for the rural poor. Unasylva 41(161): 39-52.

Charles, D. 1997 Can Bison Replace Engines of War? New Scientist 154(10 May—2081): 10.

Charles, D. 2002 Fields of dreams. New Scientist 173 (5 January-2324): 25-27.

Charpentier, P. 1902 A Comprehensive Study of Wood in all its Properties. Scott, Greenwood and Son: London, UK. 437p.

Clark, D.A. and Clark, D.B. 2001 Getting into the canopy: tree height growth in a neotropical rain forest. Ecology 82(5): 1460-1472.

Clark, J.S., Merkt, J. and Muller, H. 1989 Post-glacial fire, vegetation and human history on the northern alpine foreland, south-western, Germany. Journal of Ecology 77(4): 897-925.

Clason, T. and Robinson, J.L. 2000 From a pine forest to a silviculture system. Agroforestry Notes, No. 18. United States Department of Agriculture Forest Service, National Agroforestry Center, Lincoln, NE, USA 4p.

Cleveland C.C., Townsend, A.R., Schmidt, S.K. and Constance, B.C. 2003 Soil microbial dynamics and biogeochemistry in tropical forests and pastures of southwestern Costa Rica. Ecological Applications 13(2): 314-326.

Coates, K.D. and Burton, P.J. 1997 A gap-based approach for development of silvicultural systems to address ecosystem management objectives. Forest Ecology and Management 99: 337-354.

Cobbett, W. 1825 The Woodlands or A Treatise Describing the Trees. William Cobbett: London, UK. 601 paragraphs.

Constantine, N.L., Campbell, T.A., Baughman, W.M., Harrington, T.B., Chapman, B.R. and Miller, K.V. 2004 Effects of clearcutting with corridor retention on abundance, richness, and diversity of small mammals in the coastal plains of South Carolina, USA. Forest Ecology and Management 202(1-3): 293-300.

Cooper, C.F. 1960 Changes in vegetation, structure and growth of south-western pine forests since white settlement. Ecological Monographs 30: 129-164.

Cooper, P.J.M., Leakey, R.R.B., Rao, M.R. and Reynolds, L. 1996 Agroforestry and the mitigation of land degradation in the humid and sub-humid tropics of Africa. Experimental Agriculture 32: 235-290.

Cooper, S.M. and Ginnett, T.F. (1998) Spines protect against browsing by small climbing mammals. Oecologia 113(2): 219-221.

Couto, L., Roath, R.L., Betters, D.R., Garcia, R. and Alemeida, J.C.C. 1994 Cattle and sheep in eucalypt plantations: a silvopastoral alternative in Minas Gerais, Brazil. Agroforestry Systems 28: 173-185.

Cowling, S.A., Sykes, M.T. and Bradshaw, R.H. 2001 Palaeovegetation-model comparison, climate change and tree succession in Scandinavia over the past 1500 years. Journal of Ecology 89: 227-236.

Dagar, J.C., Singh, N.T. and Singh, G. 1995 Agroforestry options for degraded and problematic soils of India. pp. 96-120. In: P. Singh, P.S. Pathak and M.M. Roy (eds.) Agroforestry Systems for Sustainable Land Use. Science Publishers: NH, USA. 283p.

Dalling, J.W., Winter, K., Nason, J.D., Hubbett, S.P., Murawski, D.A. and Hamrick, J.L. 2001 The unusual life history of *Alseis blackiana*: a shade resistant pioneer species. Ecology 82(4): 933-945.

Dauvergne, P. 2001 Loggers and Degradation in the Asia-Pacific Corporations and Environmental Management. Cambridge University Press: Cambridge, UK. 202p.

Davidson, D.W., Cook, S.C., Snelling, R.R. and Chua, T.H. 2003 Explaining the abundance of ants in lowland tropical rainforest canopies. Science 300: 969-972.

Dawkins, H.C. and Philips, M.S. 1998 Tropical Moist Silviculture and Management: A History of Success and Failure. CAB International: NY, USA. 359p.

Dawson, T.E. 1993 Hydraulic lift and water use by plants: implications for water balance, performance and plant-plant interactions. Oecologia 95: 565-574.

Day, S. 2001 Fight the blight. New Scientist. (1 September-2306): 36-39.

de Foresta, F., Basri, A. and Wiyono 1994 A very intimate agroforestry association—cassava and improved homegardens: the Mukibat technique. Agroforestry Today 6(1)12-14.

de la Fuente, M.A.S. and Marquis, R.J. 1999 The role of ant-tended extrafloral nectaries in the protection and benefit of a Neotropical rainforest tree. Oecologia 118(2): 192-202.

DeBell, D.S., Whitesell, C.D. and Crabb, T.B. 1987 Benefits of Eucalyptus-Albizia Mixtures vary by site on Hawaii Island. USDA Forest Service, Research Paper PSW-187, 6p.

DeBell, D.S., Whitesell, C.D., and Schumbert, T.H. 1985 Mixed Plantations of Eucalyptus and Leguminous Trees Enhance Biomass Production. USDA Forest Service, Research Paper PSW-175, 6p.

Dedek, W., Pape, J., Grimmer, F. and Körner, H-J. 1989 Integrated pest control in forest management-combined use of pheromone and insecticide for attracting and killing the bark beetle *Ips typographus*, II. effects of methanidophos treatment following bark penetration into the ascending sap of pheromone-baited spruce. Forest Ecology and Management 26: 63-76.

DeMazancourt, C., Loreau, M. and Abbadie, L. 1998 Grazing optimization and nutrient cycling: when do herbivores enhance plant production. Ecology 79(7): 2242-2252.

Dix, M.E., Hill, D., Buck, L.E. and Rietveld, W.J. 1997 Forest farming: an agroforestry practice. Agroforestry Notes, No. 7. United States Department of Agriculture Forest Service, Rocky Mountain Research Station, 4p.

Donovan, T.M. and Flather, C.H. 2002 Relationships among North American songbird trends, habitat fragmentation, and landscape occupancy. Ecological Applications 12(2): 346-374.

Dorado, M., Astin, E., Verzino, G., DiRienzo, J. and Perpiñal, E. 1997 Growth curves for *Pinus elliotti, Pinus taeda,* and *Pinus radiata* in two areas of Calamuchita Valley (Córdoba, Argentina). Forest Ecology and Management 95: 173-181.

Duncan, R.S. and Chapman, C.A. 2003 Consequences of plantation harvest during tropical forest restoration in Uganda. Forest Ecology and Management 173(1-3): 235-250.

Dyrness, C.T. 1973 Early stages of plant succession following logging and burning in the western Cascade of Oregon. Ecology 54: 57-69.

Eason, W.R. 1991 The effect of leaf litter on sward botanical composition and growth. Forest Ecology and Management 45: 165-172.

Eliot, J.C. 2003 Bats bring the rainforest back. National Geographic 203(4): xiv.

Elvin, M. 2004 The Retreat of the Elephants: An Environmental History of China. Yale University Press: New Haven, CT, 564 pp.

Emboden, W.A. 1974 Bizarre Plants, Magical, Monstrous, Mythical. Macmillan Publishing Co.: NY, USA.

Emerson, G. 1846 Report on the Trees and Shrubs Growing Naturally in the Forests of Massachusetts. Dutton and Wentworth: Boston, USA. 547p.

Evans, J. 1992 Plantation Forestry in the Tropics—Tree Plantings for Industrial, Social, Environmental and Agroforestry Purpose. 2nd Edition, Clarendon Press: Oxford, UK.

Evelyn, J. 1664 Silva. A. Ward: York, UK. 649p.

Fairhead, J. and Leach, M. 1998 Reframing Deforestation-Global Analysis and Local Realities: Studies in West Africa. Routledge: London, UK. 235p.

Faunt, K. and Williams, M.R. 1997 Factors affecting the abundance of hollows in logs in jarrah forest of south-western Australia. Forest Ecology and Management 95: 153-160.

Finkeldey, R. and Ziehe, M. 2004 Genetic implications of silvicultural regimes. Forest Ecology and Management 197(1-3): 231-244.

Fish, S.K. 2000 Hohokam Impacts on Sonoran Desert Environment. pp. 250-280. In: D.L. Lentz, (ed.) Imperfect Balance Landscape Transformations in the Precolumbian Americas. Columbia University Press: NY, USA. 547p.

Fleming, T.H. 1973 Number of mammal species in North and Central American Forest Communities. Ecology 54: 555-563.

Forman, R.T.T. and Hahn, D.C. 1980 Spatial patterns of trees in a Caribbean semi-evergreen forest. Ecology 61(6): 1267-1274.

Forsey, E.S. and Braggs, E.M. 2001 Winter activity of mammals in riparian zones and adjacent forests prior to and following clear-cutting at Cooper Lake, Newfoundland, Canada. Forest Ecology and Management 145: 163-171.

Foster, D.R. 1988 Disturbance history, community organization and vegetative dynamics of the old-growth Pisgah forest, south-western New Hampshire, U.S.A. Journal of Ecology 76: 105-134.

Foster, D.R. and Zebryk, T.M. 1993 Long-term vegetation dynamics and disturbance history of a *Tsuga*-dominated forest in New England. Ecology 74(4): 982-998.

Foster, D.R., Zebryk, Y., Schoonmaker, P. and Lezberg, A. 1992 Post-settlement history of human land-use and vegetation dynamics of a *Tsuga canadensis* (hemlock) woodlot in Central New England. Journal of Ecology 80: 773-786.

Fox, J.E.D. 1968 Logging damage and influence of climber cutting prior to logging in the lowland diptercarp forest of Sabah. Malaysian Forester 31: 326-347 (referenced in Panayotou and Ashton, 1992).

Fox, T.R. 2000 Sustained productivity in intensively managed forest plantations. Forest Ecology and Management 138: 187-202.

Francis, D.R. and Foster, D.R. 2001 Response of small New England ponds to historic land use. The Holocene 11(3): 301-312.

Fredericksen, T.S. and Pariona, W. 2002 Effects of skidder disturbance on commercial tree regeneration in logging gaps in a Bolivian Tropical forest. Forest Ecology and Management 171(3): 223-230.

Frelich, L.E. and Reich, P.B. 1995 Spatial patterns and succession in a Minnesota southern-boreal forest. Ecological Monographs 65(3): 325-346.

Frost, R. 1915 North of Boston. Henry Holt: NY, USA. 137p.

Gamero, E.M., Lok, R. and Somarriba, E. 1996 Análisis agroecológico de huertos caseros traditionales en Nicaragua. Agroforestería en las Américas 3(11-12): 36-40.

Garbaye, J., Delwaulle, J.C. and Duangna, D. 1988 Growth response in eucalyptus in the Congo to ectomycorrhizal inoculation. Forest Ecology and Management 24: 151-157.

Geldenhuys, C.J. 1997 Native forest regeneration in pine and eucalypt plantation in Northern Province, South Africa. Forest Ecology and Management 99: 101-115.

Gent, F.W. 1681 Systema Agriculturæ: The Mystery of Husbandry Discovered. T.Dring: London, U.K.

Gentry, A.H. and Vasquez, R. 1988 Where have all the *Ceibas* gone? a case history of mismanagement of tropical forest resource. Forest Ecology and Management 23: 73-76.

Gibbons, P. and Lindenmayer, D.B. 1996 Issues associated with the retention of hollow-bearing trees with eucalyptus forests managed for wood production. Forest Ecology and Management 83: 245-279.

Giesser, H. and Reyer, H-U. 2004 Efficacy of hunting, feeding and fencing to reduce crop damage by wild boar. The Journal of Wildlife Management 68(4): 939-946.

Gill, E.K. and Eason, W.R. 1994 Tree protection options in silvopastoral agroforestry. Agroforestry Forum 5(2): 32-33.

Glausiusz, J. 2003 A green renaissance for the Sahel. Discover 24(1): 13.

Godoy, R. and Bennett, C.P.A. 1991 The economics of monocropping and intercropping by small holders: the case of coconuts in Indonesia. Human Ecology 19(1): 83-98.

GoI 1910 A Concise Manual of Silviculture for the Use of Forestry Students in India. Government of India, Government Printing: Calcutta, India. 186p.

Gold, M.V. 1994 Sustainable Agriculture: Definations and Terms. Special Reference SRB 94-05, National Agricultural Library: Beltsville, Maryland, USA. 10p.

Gómez-Aparicio, L., Zamora, R., Gómez, J.M., Hódar, J.A., Castro, J. and Baraza, E. 2004 Applying plant facilitation to forest restoration: a meta-analysis of the use of shrubs as nurse plants. Ecological Applications 14(4): 1128-1138.

Goodland, R.J. and Irwin, H.S. 1977 Amazonian forest and cerrado: development and environmental conservation. pp. 214-233. In: G.T. Prance, and T.S. Elias, (eds.) Extinction is Forever. New York Botanical Garden: NY, USA. 437p.

Gould, S.J. 1989 Wonderful Life: The Burgher's Shale and the Nature of History. W.W. Norton and Co.: NY, USA. 276p.

Grant, C.D. and Loneragan, W.A. 2001 The effects of burning on understory composition of rehabilitated bauxite mines in Western Australia: community change and vegetative succession. Forest Ecology and Management 145: 255-279.

Graves, H.S. 1911 The Principles of Woodlands. John Wiley and Sons: NY, USA. 320p.

Gravitz, L. 2002 Coffee with a killer. Discover 22(12): 15.

Gravitz, L. 2003 Salvage the trees, spoil the forest. Discover 24(4): 15.

Green, S.B. 1908 Principles of American Forestry. John Wiley and Sons: NY, USA. 334p.

Greene, D.F. and Johnson, E.A. 1996 Wind dispersal of seeds from a forest into a clearing. Ecology 77(2): 595-609.

Groninger, J.W., Zedaker, S.M. and Fredericksen, T.S. 1997 Stand characteristics of inter-cropped loblolly pine and black locust. Forest Ecology and Management 91: 221-227.

Gupta, G.N., Mutha, S. and Limba, N.K. 2000 Growth of *Albizia lebbeck* on micro-catchments in the Indian arid zone. International Tree Crops Journal 10: 193-202.

Gustafson, F.J. 1996 Expanding the scale of forest management: allocating timber harvests in time and space. Forest Ecology and Management 87: 27-39.

Harrington, R.A. and Ewel, J.J. 1997 Invasibility of tree plantations by native and non-indigenous plant species in Hawaii. Forest Ecology and Management 99: 153-162.

Hartshorn, G.S. 1989 Application of gap theory to tropical forest management: natural regeneration on strip clear-cuts in the Peruvian Amazon. Ecology 70(3): 567-569.

Hauser, S., Vanlauwe, B., Asawalam, D.O. and Norgrove, L. 1997 Role of earthworms in traditional and improved low-input agricultural systems in West Africa. pp. 113-136. In: L. Brussaard and R. Ferrara-Cerrato (eds.) Soil Ecology in Sustainable Agricultural Systems. Lewis Publishers: NY, USA. 168p.

Hawley, R.C. 1921 The Practice of Silviculture with Particular Reference to its Application in the United States. John Wiley and Sons: NY, USA.

Hoffmann, R.R. 1973 The Ruminant Stomach: Stomach Structure and Feeding Habit of East African Game Ruminants. East African Literature Bureau: Nairobi, Kenya (referenced in Vera, 2000).

Holmgren, P., Masakha, E.J. and Sjöholm, H. 1994 Not all African land is being degraded: a recent survey of trees on farms in Kenya reveals rapidly increasing forest resources. Ambio 23(7): 390-395.

Houston, D.B. 1973 Wildfires in the northern Yellowstone National Park. Ecology 54(5): 1111-1117.

Hsiung (Xiong), W., Yang, S. and Tao, S. 1995 Historical development of agroforests in China. Agroforestry Systems 30: 277-287.

Huddle, J.A. and Pallardy, S.G. 1996 Effects of long-term annual and periodic burning on tree survival and on a Missouri Ozak oak-hickory forest. Forest Ecology and Management 82: 1-9.

Hughes, J.D. 1994 Pan's Travail: Environmental Problems of the Ancient Greeks and Romans. John Hopkins University Press: Baltimore, USA. 277p.

Hulugalle, N.R. and Ezumah, H.C. 1991 Effects of cassava-based cropping systems on physio-chemical properties of soil and earthworm casts in a tropical alfisol. Agriculture, Ecosystems and Environment 25: 55-63.

Hutchins, H.E., Hutchins, S.A. and Lin, B-W. 1996 The role of birds and mammals in Korean pine (*Pinus koraiensis*) regeneration dynamics. Oecologia 107(1): 120-130.

Huxley, P.A. 2001 Multipurpose trees: biological and ecological aspects relevant to their selection and use. pp. 19-74. In: F.T. Last, (ed.) Tree Crop Ecosystems. Elsevier: Amsterdam, 492p.

Jackson, S.M., Fredericksen, T.S. and Malcolm, J.R. 2002 Area disturbed and residual stand damage following logging in a Bolivian tropical forest. Forest Ecology and Management 166(1-3): 271-283.

Jain, S.K. 2000 Human aspects of plant diversity. Economic Botany 54(4): 459-470.

Jensen, M. 1993 Soil conditions, vegetation structure and biomass of a Javanese homegarden. Agroforestry Systems 24: 171-186.

Johns, J.S., Barreto, P. and Uhl, C. 1996 Logging damage during planned and unplanned logging operation in the eastern Amazon. Forest Ecology and Management 89: 57-77.

Kalkhoven, J.T.R. 1993 Survival of populations and the scale of the fragmented agricultural landscape. pp. 83-90. In: R.G.H. Bunce, L. Ryszkowski and M.G. Paoletti (eds.) Landscape Ecology and Agroecosystems. Lewis Publishers: London, UK. 241p.

Kardill, L. 1980 Forest berries and mushrooms-an endangered resource. Ambio 6(5): 241-247.

Kareiva, P. 1994 Diversity begets productivity. Nature 368(21 April—6473): 686-687.

Kattan, G.H. and Alvarez-López, H. 1996 Preservation and management of biodiversity in fragmented landscapes in the Colombia Andes. pp. 3-18. In: J. Schelhas and R. Greenberg (eds.) Forest Patches in Tropical Landscapes. Island Press: Washington, DC, USA. 426p.

Kelty, M.J. 1992 Comparative productivity of monocultures and mixed-species stands. pp. 125-141. In: M.J. Kelty, B.C. Larson and C.D. Oliver (eds.) The Ecology and Silviculture of Mixed Species Forests. Kluwer Academic Publishers: The Netherlands, 287p.

Kidundo, M. 1997 *Melia volkensii*—propagating the tree of knowledge. Agroforestry Today 9(2): 21-22.

Kienast, F., Fritschi, J., Bissegger, M. and Abderholden, W. 1999 Modeling successional patterns of high-elevation forests under changing herbivore pressure-responses at the landscape level. Forest Ecology and Management 120: 35-46.

Kilgore, B.M. and Taylor, D. 1979 Fire history of a sequoia-mixed conifer forest. Ecology 60(1): 129-142.

King, D.I. and DeGraaf, R.M. 2000 Bird species diversity and nesting success in mature, clearcut, and shelterwood forest in northern New Hampshire. Forest Ecology and Management 129: 227-235.

Kittredge, D.B. and Parker, M. 1999 Massachusetts Forestry—Best Management Practices. 3rd Printing, Department of Environmental Management: Boston, USA. 56p.

Kleine, M. and Heuveldop, J. 1993 A management planning concept for sustainable yield of tropical forests in Sabah, Malaysia. Forest Ecology and Management 61: 277-297.

Klironomos, J.N. 2002 Another Form of Bias in Conservation Research. Science 298(25 October—5594): 749.

Kneeshaw, D.D. and Bergeron, Y. 1998 Canopy gaps and tree replacement in the southeastern boreal forest. Ecology 79(3): 783-794.

Koch, P. 1996 Lodgepole Pine in North America. Vol. 1, Forest Products Society: Madison, WI, USA. 342p.

Koech, E.K. and Whitbread, R. 2000 Disease incidence and severity in beans in alleys between leucaena hedgerows in Kenya. Agroforestry Systems 49: 85-101.

Kumar, B.M., Kumar, S.S. and Fisher, R.F. 1998 Intercropping teak with *Leucaena* increases tree growth and modifies soil characteristics. Agroforestry Systems 42: 81-89.

Laiolo, P., Caprio, E. and Rolando, A. 2003 Effects of logging and non-native proliferation on the birds overwintering in the upland forests of north-western Italy. Forest Ecology and Management 179(1-3): 441-454.

Lamb, F.B. 1966 Mahogany of Tropical America: Its Ecology and Management. University of Michigan Press: Ann Arbor, USA. 220p.

Langston, N. 1995 Forest Dreams, Forest Nightmares. University of Washington Press: Seattle, WA, USA. 367p.

Leach, M. 1994 Rainforest Relations, Gender, and Resource Use Among the Mende of Gola, Sierra Leone. Smithsonian Institute Press: Washington, DC, USA. 272p.

Ledgard, N. 2001 The spread of lodgepole pine (*Pinus contorta*, Dougl.) in New Zealand. Forest Ecology and Management 141: 43-57.

Lefroy, E.C., Hobbs R.J., Connor, M.H. and Pate, J.S. 1999 What can agroforestry learn from natural ecosystems? Agroforestry Systems 45: 423-436.

Leuschner, C., Hertel, D., Coners, H. and Büttner, V. 2001 Root competition between beech and oak: a hypothesis. Oecologia 126: 176-184.

Levey, D.J. 1988 Tropical wet forests, treefall gaps and distributions of understory birds and plants. Ecology 69(4): 1076-1089.

Libbie, A.R. and Freudenberger, M.S. 1996 Sacred groves in Africa: forest patches in transition. pp. 300-324. In: J. Schelhas and R. Greenberg (eds.) Forest Patches in Tropical Landscapes. Island Press: Washington, DC, USA. 426p.

Liebold, M.A. 1995 Mechanistic models and community context. Ecology 76(5): 1371-1382.

Linares, O.F. 1976 Garden hunting in the American tropics. Human Ecology 4(4): 331-349.

Lloyd, A.H. and Graumlich, L.J. 1997 Holocene dynamics of treeline forests in the Sierra Nevada. Ecology 78(4): 1199-1210.

Lockaby, B.G., Stanturf, J.A. and Messina, M.G. 1997 Effects of silvicultural activity on ecological processes in floodplain forests of the southern United States: a review of existing reports. Forest Ecology and Management 90: 93-100.

Loumeto, J.J. and Huttel, C. 1997 Understory vegetation in fast-growing tree plantation on savanna soils in Congo. Forest Ecology and Management 99: 65-81.

Lovett, P.N. and Naq, N. 2000 Evidence for the anthropic selection of the sheanut tree (Vitellaria paradoxa). Agroforestry Systems 48: 273-288.

Lowe, R.G. and Ugbechie, F.N. 1975 Kennedy's Natural Regenerative Experiment at Sapoba, after 45 Years. Unpublished Report to the Forestry Association of Nigeria, 6th Annual Conference, Calabar, Nigeria (referenced in Panayotou and Ashton, 1992).

Lugo, A.E. 1997 The apparent paradox of reestablishing species richness on degraded lands with tree monocultures. Forest Ecology and Management 99: 9-19.

Lyon, J. and Herwich, R.H. 1996 Modification of tropical forest patches for wildlife protection and community conservation in Belize. pp. 205-230. In: J. Schelhas and R. Greenberg (eds.) Forest Patches in Tropical Landscapes. Island Press: Washington, DC, USA. 426p.

MacKenze, D. 1999 Good ol' milkweed. New Scientist 164(16 October-2208): 24.

MacLellan, C.R. 1970 Woodpecker ecology in the apple orchard environment. pp. 273-284. In: Proceedings of the Tall Timbers Conference on Ecological Animal Control by Habitat Management. Vol. 2, Feb. 26-28 1970, Tall Timbers Research Station: Tallahassee, FL, USA. 286p.

Magusson, W.E., de Lima, O.P., Reis, F.Q., Higuchi, N. and Ramos, J.F. 1999 Logging activity and tree regeneration in a Amazonian Forest. Forest Ecology and Management 113: 64-74.

Malarson, G.P. and Trabaud, L. 1988 Vigour of post-fire resprouting by *Quercus coccifera* L. Journal of Ecology 76(2): 351-365.

Malcolm, J.R. 1994 Edge effects in central Amazonian forest fragments. Ecology 75(8): 2438-2445.

Mando, A. and van Rheenen, T. 1998 Termites and agricultural production in the Sahel: from enemy to friend. Netherlands Journal of Agricultural Science 64: 77-85.

Mann, C.C. 2002a 1491. The Atlantic Monthly. 289(3): 41-52.

Mann, C.C. 2002b The real dirt on rainforest fertility. Science 297(9 August-5583): 920-923.

Marc, P., Canard, A. and Ysnel, F. 1999 Spiders (Araneae) useful for pest limitation and bioindication. Agriculture, Ecosystems and Environment 74: 229-273.

Marn, H.M. and Jonkers, W.P. 1981 Logging Damage in Tropical High Forest. Presented Paper; International Forestry Seminar, Kuala Lumpur, Malaysia (referenced in Panayotou and Ashton, 1992).

Matlack, G.R. 1994 Plant species migration in mixed-history forest landscapes in eastern North America. Ecology 75(5): 1491-1502.

Matthews, J.D. 1989 Silvicultural Systems. Clarendon Press: Oxford, UK. 284p.

Mazurek, M.J. and Zielinski, W. 2004 Individual legacy trees influence vertebrate wildlife diversity in commercial forests. Forest Ecology and Management 193(3): 321-324.

McNeely, J.A. 1993 Economic incentives for conserving biodiversity: lesson from Africa. Ambio 22(2-3): 144-150.

Méndez, V.E., Lok, R. and Somarriba, E. 2001 Interdisciplinary analysis of homegardens of Nicaragua: micro-zonation, plant use and socio-economic importance. Agroforestry Systems 51: 85-96.

Menzies, N. 1988 Three hundred years of taungya: a sustainable system of forestry in South China. Human Ecology 16(4): 361-376.

Mesquita, R. de C. G. 2000 Management of advanced regeneration in secondary forests of the Brazilian Amazon. Forest Ecology and Management 130: 131-140.

Meyers, R.K. and van Lear, D.H. 1998 Hurricane-fire interactions in coastal forests of the south: a review and hypothesis. Forest Ecology and Management 103: 265-276.

Michon, G. and DeForesta, H. 1997 Agroforests: pre-domestication of forest trees or tree domestication of forest ecosystems. Netherlands Journal of Agricultural Science 45: 451-462.

Milius, S. 2003a Any hope for old chestnuts? Science News 163(18): 282-284.

Milius, S. 2003b Sweet Lurkers. Science News 164(13 December-24): 374.

Miller, D.R., Rosenberg, N.J. and Bagley, W.T. 1973 Soybean water use in the shelter of a slat-fence windbreak. Agricultural Meteorology 11: 405-418.

Miller, R.E. and Murray, M.D. 1978 The effects of red alder on growth of douglas-fir. pp. 283-306. In: D.G. Briggs, D.S. DeBell and W.A. Atkinson (eds.) Utilization and Management of Alder. Proceedings of a Symposium, April 25-27, 1977, USDA: Portland, Oregon, USA. 283p.

Mohd Ali, A.R. and Raja Bariza, R.S. 2001 Intercropping rattan with rubber and other crops. Unasylva 52(205): 9-11.

Molles, M.C. 1999 Ecology: Concepts and Applications. McGraw Hill: NY, USA. 509p.

Molofsky, J. and Angspurger, C.K. 1992 The effects of leaf litter on early seedling establishment in a tropical forest. Ecology 73(1): 68-77.

Moon, F. and Brown, N.C. 1914 Elements of Forestry. John Wiley and Sons: NY, USA. 409p.

Moreno, D., Croce, G.F., Guido, M.A. and Montanari, C. 1993 Pine plantations on ancient grasslands: ecological change in the Mediterranean mountains of Ligura, Italy during the 19th and 20th centuries. pp. 93-110. In: C. Watkins (ed.) Ecological Effects of Afforestation in the History and Ecology of Afforestation in Western Europe. CAB International: NY, USA. 224p.

Munishi, P.K.T. and Chamshama, S.A.O. 1994 A study of wind damage on *Pinus patula* stands in southern Tanzania. Forest Ecology and Management 63: 13-21.

Naeem, S., Thompson, L.J., Lawler, S.P., Lawton, J.H. and Woodfin, R.M. 1994 Declining biodiversity can alter the performance of ecosystems. Nature 368(21 April-6473): 734-735.

Nicholson, D.I. 1958 An analysis of logging damage in tropical rain forest, North Borneo. Malaysian Forester 29: 182-203 (referenced in Panayotou and Ashton, 1992).

Nisbet, J. 1905 The Forester. William Blackwood and Sons: London, UK, Vol. 1, 506p.

O'Conner, R.J. and Shrubb, M. 1986 Farming and Birds. Cambridge University Press: NY, USA. 290p.

Ogden, J. and Stewart, G.H. 1995 Community dynamics of the New Zealand conifers. pp. 81-119. In: N.S. Enright and R.S. Hill (eds.) Ecology of the Southern Conifers. Smithsonian Institute Press: Washington, DC, USA. 342p.

Ohte, N., Koba, K., Yoshikawa, A., Sugimoto, A., Matsuo, N., Kabeya, N. and Wang, L. 2003 Water utilization of natural and planted trees in the semi-arid desert of Inner Mongolia, China. Ecological Applications 13(2): 337-351.

Palm, C.A. 1995 Contribution of agroforestry trees to nutrient requirements of intercropped plants. Agroforestry Systems 30: 105-124.

Panayotou, T. and Ashton, P.S. 1992 Not by Timber Alone: Economics and Ecology for Sustaining Tropical Forests. Island Press: Washington, DC, USA. 282p.

Pearce, F. 2003a A grayer shade of green. The New Scientist 178(21 June-2400): 41-43.

Pearce, F. 2003b Fragmented forest faces bleak future. The New Scientist 180(4 November-2420): 14-15.

Pearce, F. 2004 The accidental rainforest. The New Scientist 183(18 September-2465): 44-45.

Penalba, M.C. 1994 The history of holocene vegetation in northern Spain from pollen analysis. Journal of Ecology 82(4): 815-832.

Pennisi, E. 2003 Fungi shield new host plants from heat and drought. Science 301(12 September-5639)1466.

Pennisi, E. 2004 The secret life of fungi. Science 304(11 June-5677): 1620-1622.

Perfecto, I., Rice, R.A., Greenberg, R. and van der Voort, M.E. 1996 Shade coffee: a disappearing refuge for biodiversity. BioScience 46(8): 598-608.

Perry, C.H., Miller, R.C. and Brooks, K.N. 2001 Impacts of short-rotation hybrid poplar plantation on regional water yield. Forest Ecology and Management 143: 143-151.

Peterken, G.F. 1993 Long-term floristic development of woodland on former agricultural lands in Lincolnshire, England. pp. 31-43. In: C. Watkins (ed.) Ecological Effects of Afforestation: Studies in the History and Ecology of Afforestation in Western Europe. CAB International: NY, USA. 224p.

Peterken, G.F. 2001 Ecological effects of introduced tree-species in Britain. Forest Ecology and Management 141: 31-42.

Peterson, C.J. and Squiers, E.R. 1995 Competition and succession in an aspen-white-pine forest. Journal of Ecology 83: 449-457.

Peterson, J.T. 1981 Game, farming, and inter-ethnic relations in northeastern Luzon, Philippines. Human Ecology 9(1): 1-22.

Petit, B. and Montagnini, F. 2004 Growth equations and rotation ages of ten native tree-species in mixed and pure plantations in the humid neotropics. Forest Ecology and Management 199(2-3): 243-257.

Pitman, N.C.A., Terborgh, J., Silman, M.R. and Nuñez V., P. 1999 Tree-species distribution in an upper Amazonia forest. Ecology 80(8): 2651-2661.

Platt, W.J. and Connell, J.H. 2003 Natural disturbance and directional replacement of species. Ecological Monographs 73(4): 507-522.

Pontey, W. 1805 The Forest Pruner. White and Mawman: London, UK. 277p.

Preisser, E.L. 2003 Field evidence for a rapidly cascading underground food web. Ecology 84(4): 869-874.

Putz, F.E. 1983. Treefall pits and mounds, buried seeds and the importance of soil disturbance to pioneer trees in Barro Colorado Island, Panama. Ecology 64(5): 1069-1074.

Pye-Smith, C. 2003 Fruits of the Forest. New Scientist 179(19 July-2404): 36-39.

Rabin, M. and Thaler, R.H. 2001 Anomalies: risk adversion. Journal of Economic Perspective. 15(1): 219-232.

Rackham, O. 1998 Savanna in Europe. pp. 1-24. In: K.J. Kirby, and C. Watkins, (eds.) The Ecological Basis of European Forests. CAB International: NY, USA. 373p.

Rama Rao, M. 1911 Host plants of the sandal tree. The Indian Forest Records 11(iv): 159-207.

Rao, A.V. and Giller, K.E. 1993 Nitrogen fixation and its transfer from *Leucaena* to grass using ^{15}N. Forest Ecology and Management. 61: 221-227.

Raynor, W. 1992 Economic analysis of indigenous agroforestry: a case study on Pohnpei Island, Federated State of Micronesia. pp. 243-258. In: G.H. Sullivan, S.M. Huke and J.M. Fox (eds.) Financial and Economic Analysis of Agroforestry Systems: Proceedings of a Workshop held in Honolulu, Hawaii, USA, July 1991. Nitrogen Fixing Tree Association: Paia, HI, USA. 312p.

Redford, K.H., Klein, B. and Murcia, C. 1992 Incorporation of game animals into small-scale agroforestry systems in the neotropics. pp. 334-346. In: K.H. Redford and C. Padoch (eds.) Conservation of Neotropical Forests. Columbia University Press: NY, USA. 475p.

Rehfeldt, G.E. 1978 Genetic differentiation of douglas-fir population from the northern rocky mountain. Ecology 59(6): 1264-1270.

Revkin, A.C. 2000 Now preventing forest fires: smokey goats. The New York Times. June 13: F3.

Reynolds, H.L., Packer, A., Bever, J.D. and Clay, K. 2003 Grass roots ecology: plant-microbe-soil interactions as drivers of plant community structure and dynamics. Ecology 84(9): 2281-2291.

Rice, R.E., Gullison, R.E. and Reid, J.W. 1997(a) Can sustainable management save tropical forests? Scientific American 276(4): 44-49.

Rice, R.E., Gullison, R.E. and Reid, J.W. 1997b Vive la différence. Scientific American 276(4): 48.

Risch, S.J. and Carroll, C.R. 1982 Effect of a predaceous ant, *Solenopis geminata*, on arthropods in a tropical agroecosystem. Ecology 63: 1979-1983.

Robinson, J.G. 1996 Hunting wildlife in forest patches: an ephemeral resource. pp. 111-130. In: J. Schelhas and R. Greenberg (eds.) Forest Patches in Tropical Landscapes. Island Press: Washington, DC, USA. 426p.

Roth, F. 1902 The First Book of Forestry. Ginn and Co.: Boston, USA. 291p.

Ruc, J. 1990 Compounds from plants that regulate or participate in disease resistance. pp. 213-224. In: D.J. Chadwick and J. March (eds) Bioactive Compounds from Plants, John Wiley and Sons, NY, USA. 242p.

SAF 1944 Forest Terminology: A Glossary of Technical Terms used in Forestry. Society of American Foresters: Washington, DC, USA.

Salafsky, N. 1994 Forest gardens in the Gunung Palung region of West Kalimantan, Indonesia. Agroforestry Systems 28: 237-268.

Sandström, K. 1998 Can forests "provide" water: widespread myth or scientific reality. Ambio 27(2): 132-138.

Santora, A.E., Lombardero, M.J., Ayres, M.P. and Ruel, J.J. 2001 Interactions between fire and bark beetles in an old growth pine forest. Forest Ecology and Management 144: 245-254.

Schenck, C.A. 1904 Forest Utilization, Mensuration and Silviculture. Biltmore?: NC, 3parts.

Schlich, W. 1910 Schlich's Manual of Forestry. Vol. II, 4th Edition. Bradbusy, Agnew and Co.: London, UK. 424p.

Schmiegelow, F.K.A., Machtans, C.S. and Hannon, S.J. 1997 Are boreal birds resilient to forest fragmentation? An experimental study of short-term community response. Ecology 78(6): 1914-1932.

Schnitzer, S.A. and Carson, W.P. 2001 Treefall gaps and the maintenance of species diversity in a tropical forests. Ecology 82(4): 913-919.

Schroth, G. 1995 Tree root characteristics: a criterion for species selection and system design in agroforestry. Agroforestry Systems 30: 125-143.

Schroth, G., Krauss, V., Gasparotto, L., Duarte Aguilar, J.A. and Vohland, K. 2000 Pests and diseases in agroforestry systems of the tropics. Agroforestry Systems 50: 199-241.

Schwartz, M.W., Brigham, C.A., Hoeksema, J.D., Lyons, K.G., Mills, M.H. and van Mantgen, P.J. 2000 Linking biodiversity to ecosystem function: indications for conservation ecology. Oecologia 122: 297-305.

Setälä, H. 2000 Reciprocal interactions between Scots pine and soil web structure in the presence and absence of ectomycorrhiza. Oecologia 125: 109-118.

Seydack, A.H.W. 1995 An unconventional approach to timber yield regulation for multi-aged, multi-species forest I. fundamental consideration. Forest Ecology and Management 77: 139-153.

Seydack, A.H.W., Vermeulen, W.J., Heyns, H.E., Durrheim, G.P., Vermeulen, C., Williams, D., Ferguson, M.A., Huisamen, J. and Roth, J. 1995 An unconventional approach to timber yield regulation for multi-aged, multi-species forest II. Application to a South African forest. Forest Ecology and Management 77: 155-168.

Sharma, P.N. and Silva, A.D.S. 1987 Native forest (caatinga) watershed management for runoff inducement for irrigation. Forest Ecology and Management 18: 73-84.

Shaver, P. 2002 Fire and silviculture. Inside Agroforestry. National Agroforestry Center, USDA: Nebraska, USA. Fall: 1.

Shrair, A.J. 2000 Agricultural intensity and its measure in frontier regions. Agroforestry Systems 49: 301-318.

Silva, V., Bandyopadhyay, J. and Jayal, N.D. 1985 Afforestation in India: problems and strategies. Ambio 14(6): 329-333.

Simberoff, D. 1999 The role of science in the preservation of forest biodiversity. Forest Ecology and Management 115(2-3): 101-111.

Simon, K-H. (ed.) 2001 Lexicon Silvestre, Secunda Pars, Dictionary of Forest Affairs. First Edition, Förderverein "Lexicon Silvestre", e.V.: Eberswalde, Germany.

Simpson, J. 1903 The New Forestry. Pawson and Brailsford: Sheffield, UK. 220p.

Sist, P., Nolan, T., Bertault, J-G. and Dykstra, D. 1998 Harvesting intensity versus sustainability in Indonesia. Forest Ecology and Management 108: 251-260.

Smith, J. 1870 Treatise on the Management and Cultivation of Forest Trees. Blackie and Son: London, UK. 164p.

Sollins, P. 1998 Factors influencing species composition in tropical lowland rain forest: does soil matter? Ecology 79(1): 23-30.

Soluri, J. 2001 Altered landscapes and transformed livelihoods: banana companies, Panama disease, and rural communities on the north coast of Honduras. pp. 15-30. In: C. Flora (ed.) Interactions between Agroecosystems and Rural Communities. CRC Press: Boca Ratan, FL, USA. 273p.

Southgate, D. 1998 Tropical Forest Conservation: An Economic Assessment of the Alternative in Latin America. Oxford University Press: Oxford, U.K. 175p.

Stanton, M. and Young, T. 1999 Thorny relationships. Natural History 108(9): 28-31.

Stapley, L. 1998 The interaction of thorns and symbiotic ants as an effective defence mechanism of swollen-thorn acacias. Oecologia 115(3): 401-405.

Statälä, H. and Huhta, V. 1991 Soil fauna increase *Betula pendula* growth: laboratory experiments with coniferous forest floor. Ecology 72(2): 665-671.

Steenbergh, W.F. and Lowe, C.H. 1969 Critical factors during the first year of life of the saguaro (*Cereus giganteus*) at Saguaro Monument, Arizona. Ecology 50: 823-834.

Stephens, S.L., Dulitz, D.J. and Mertin, R.F. 1999 Giant sequoia regeneration in group selection openings in the southern Sierra Nevada. Forest Ecology and Management 120: 89-95.

Sterner, R.W., Ribic, C.A. and Schatz, G.E. 1986 Testing for life historical changes in spatial patterns of four tropical tree-species. Journal of Ecology 74(3): 621-633.

Stokstad, E. 2003 'Pristine' forests teemed with people. Science 301(19 September—5640): 1645-1646.

Stratton, L.C., Goldstein, G. and Meinzer, F.C. 2000 Temporal and spatial partitioning of water resources among eight woody species in a Hawaiian dry forest. Oecologia 124: 309-317.

Stringer, W.C. and Alverson, D.R. 1994 Use of fire as a management tool in alfalfa production ecosystems. pp. 492-493. In: K.L. Cambell, W.D. Graham and A.B. Bottcher (eds.) Environmentally Sound Agriculture. Proceedings of the second conference, April 20-22 1994, Orlando, FL, USA. 577p.

Strout, B.B., Deschenes, J.M. and Oharmann, L.F. 1975 Multi-species model of a directional forest. Ecology 56: 226-231.

Struhsaker, T.T., Kasenene, J.M., Gaither Jr., J.E., Larsen, N., Musango, S. and Barcroft, R. 1989 Tree mortality in Kibali forest, Uganda: a case study of dieback in a tropical rain forest adjacent to exotic conifer plantations. Forest Ecology and Management 29: 165-185.

Suckling, D.M. and Karg, G. 1998 Pheromones and other semichemicals. pp. 63-99. In: J.E. Rechcigl and N.A. Recheigl (eds.) Biological and Biotechnological Control of Insect Pests. Lewis Publishers, Boca Raton, FL, USA. 374 pp.

Svenning, J-C., Kinner, D.A., Stallard, R.F., Engelbrecht, B.M.J. and Wright, S.J. 2004 Ecological determinism in plant comunity structure across a tropical forest landscape. Ecology 85(9): 2526-2538.

Szott, L.T., Fernandes, E.C.M. and Sanchez, P.A. 1991 Soil-plant interactions in agroforestry systems. Forestry Ecology and Management 45: 127-152.

Taguder, E.T. and Reyes, M.R. 1962 Silvicultural practices in some tropical countries and their possible application in the Philippines. The Philippine Journal of Forestry 16(1-2): 47-78

Taylor, R.J. and Haseler, M.E. 1995 Effects of partial logging system on bird assemblages in Tasmania. Forest Ecology and Management 72: 131-149.

Tilki, F. and Fisher, R.F. 1998 Tropical leguminous species for acid soils: studies on plant form and growth in Costa Rica. Forest Ecology and Management 108: 175-192.

Tilman, D., Knops, J., Wedin, D., Reich, P., Ritchie, M. and Siemann, E. 1997 The influence of functional diversity and composition on ecosystem processes. Science 277(29 August-5330): 1300-1306.

Troup, R.S. 1928 Silvicultural Systems. Clarendon Press: Oxford, UK. 199p.

Troup, R.S. 1940 Colonial Forest Administration. Oxford Press: London, UK. 476p.

Tseplyaev, V.P. 1965 The Forests of the USSR. Israel Program for Scientific Translations: Jerusalem, Israel. 521p.

Tuomela, K., Kuusipalo, J., Vesa, L., Nuryanto, K., Sagala, A.P.S. and Ådjers, G. 1996 Growth of dipterocarp seedlings in artificial gaps: an experiment in a logged-over rainforest in South Kalimenta, Indonesia. Forest Ecology and Management 81: 95-100.

Turner, N.J. 1994 Burning mountainsides for better crops: aboriginal landscape burning in British Columbia. pp. 311-317. In: M. Pilarski (ed.) Restoration Forestry: An International Guide to Sustainable Forestry Practice. Kivakí Press: Durango, CO, USA 525p.

Valderrábano, J. and Torrono, L. 2000 The potential for using goats to control *Genista scorpius* shrubs in European black pine stands. Forest Ecology and Management 126: 377-383.

van der Valk, H.C., Niassy, A. and Bèye, A.B. 1999 Does grasshopper control create grasshopper problem? —monitoring side-effects of fenitrothian applications in the western Sahel. Crop Protection 18(2): 139-149.

Vandermeer, J. 1995 The ecological basis of alternative agriculture. Annual Review of Ecology and Systematics 26: 210-224.

Vera, F.W.M. 2000 Grazing Ecology and Forest History. CABI Publishing. The Netherlands 506p.

Versteeg, M.N., Amadji, F., Eteka, A., Gogan, A. and Koudokpon, V. 1998 Farmers' adoptability of *Mucuna* fallowing and agroforestry technologies in the coastal savanna of Benin. Agricultural Systems 56: 269-287.

Vieyra-Odilon, L. and Vibrans, H. 2001 Weeds as crops: the value of maize field weeds in the valley of Toluca, Mexico. Economic Botany 55(3): 426-443.

Víquez, E., Prado, A., Oñoro, P. and Solabo, R. 1994 Caracterización del Huerto mixto tropical 'la asunción', Masatepe, Nicaragua. Agroforestería en las Américas 1(2): 5-9.

Walsh, K. 1997 The chocolate bug. Natural Science 106(4): 54-57.

Walter, G.H. 2003 Insect Pest Management and Ecological Research. Cambridge University Press: Cambridge, UK. 387p.

Wang, H. 1994 Tea and trees: a good blend from China. Agroforestry Today 6(1): 6-8.

Wang, Q. and Shogren, J.F. 1992 Characteristics of the crop-paulownia system in China. Agriculture, Ecosystems and Environment 39: 145-152.

Wardle, D.A., Walker, L.R. and Bardgett, R.D. 2004 Ecsosystem properties and forest decline in contrasting long-term chronosequences. Science 305(23 July – 5683): 509-513.

Waring, R.H. and Pitman, G.B. 1985 Modifying lodgepole pine stands to change susceptibility to mountain pine beetle attack. Ecology 66(3): 884-897.

Weaver, J.C. 1919 The Ecological Relations of Roots. Publication 286, Carnegie Institute of Washington, Washington D.C., USA.

Weaver, J.C. 1920 Root Development in Grassland Formulation. Publication 292, Carnegie Institute of Washington, Washington D.C., USA.

West, N.E. 1968 Rodent-influenced establishment of ponderosa pine and bitterbrush seedlings in central Oregon. Ecology 49: 1009-1011.

White, A.S. 1985 Presettlement regeneration patterns in s southwestern ponderosa pine stand. Ecology 66(2): 584-594.

Whitman, A.A., Brokow, N.V.L. and Hogan, J.M. 1997 Forest damage caused by selective logging of mahogany (*Swietenia macrophylla*) in northern Belize. Forest Ecology and Management 92: 87-96.

Whitmore, T.C. 1975 Tropical Rainforests of the Far East. Clarendon Press: Oxford, UK. 282p.

Wiersum, K.F. 1984 Surface erosion under various tropical agroforestry systems. pp. 231-339. In: C.L. O'Loughlin and A.J. Pearce (eds.) Symposium on Effects of Forest Land-use on Erosion and Slope Stability. East-West Center: Hawaii, 310p.

Williams, M. 1989 Americans and their Forests: A Historical Geography. Cambridge University Press: Cambridge, 597p.

Williams, T. 2002 America's largest weed. Audubon 104(1): 24-31.

Williams-Linera, G. 1990 Vegetative structure and environmental conditions of forest edges in Panama. Journal of Ecology 78(2): 356-373.

Wilsey, B.J. and Potvim, C. 2000 Biodiversity and ecosystem functioning: importance of species evenness in an old field. Ecology 81(4): 887-892.

Wingfield, M.J. and Swart, W.S. 1994 Integrated management of forest tree diseases in South Africa. Forest Ecology and Management 65: 11-16.

Withner, C.L. 1977 Threatened and endangered species of orchids. pp. 314-322. In: G.T. Prance and T.S. Elias (eds.) Extinction is Forever. New York Botanical Garden: NY, USA. 437p.

Wohlgemuth, T. Bürgi, M. Scheidegger C. and Schütz, M. 2002 Dominance reduction of species through disturbance–a proposal for central European forests. Forest Ecology and Management 166(1-3): 1-15.

Wojtkowski, P.A. 1993 Toward an understanding of tropical home gardens. Agroforestry Systems 24: 215-222.

Wojtkowski, P.A. 1998 The Theory and Practice of Agroforestry Design. Science Publishers: NH, USA. 282p.

Wojtkowski, P.A. 2002 Agroecological Perspectives in Agronomy, Forestry and Agroforestry. Science Publishers: NH, USA.

Wojtkowski, P.A. 2004 Landscape Agroecology. Food Products Press: NY, USA. 375p.

Woolsey, T.S. and Greeley, W.B. 1920 Studies in French Forestry. John Wiley and Sons: NY, USA. 536p.

Worlidge, J. 1689 The Second Part of Systema Agriculturæ or the Mystery of Husbandry. F.W. Gent: London, UK.

Yamada, I. 1997 Tropical Rain Forests of Southeast Asia: A Forest Ecologist's View. University of Hawaii Press: Honolulu, USA. 392p.

Yeaton, R.L. 1988 Porcupines, fires and the dynamics of tree layer of the *Burkea africana* savanna. Journal of Ecology 76(4): 1017-1029.

Yokoyama, S., Maeji, I., Ueda, T., Ando, M. and Shibata, E. 2001 Impact of bark stripping by Sika deer *Cervus nippon* on subalpine coniferous forests in central Japan. Forest Ecology and Management 140: 93-99.

Yoon, C.K. 2000 Simple method found to increase crop yield vastly. The New York Times. August 22: D1-D2.

Yoon, C.K. 2001 Something missing in fragile cloud forest: the clouds. The New York Times. November 20: D4.

Yoshida, T. and Kamitani, T. 1997 The stand dynamics of a mixed coppice forest of shade tolerant and intermediate species. Forest Ecology and Management 95: 35-43.

Zobel, B.J., van Wyk, G. and Stahl, P. 1987 Growing Exotic Forests. John Wiley and Sons: NY, USA. 508p.

Author Index

Index

For Product Safety Concerns and Information please contact our EU representative GPSR@taylorandfrancis.com Taylor & Francis Verlag GmbH, Kaufingerstraße 24, 80331 München, Germany

Batch number: 10726382

Printed by Printforce, the Netherlands